Lecture Notes in Artificial Intelligence   8817

Subseries of Lecture Notes in Computer Science

More information about this series at http://www.springer.com/series/1244

Wei Lee Woon · Zeyar Aung
Stuart Madnick (Eds.)

Data Analytics for Renewable Energy Integration

Second ECML PKDD Workshop, DARE 2014
Nancy, France, September 19, 2014
Revised Selected Papers

 Springer

Editors
Wei Lee Woon
Masdar Institute of Science and Technology
Abu Dhabi
UAE

Stuart Madnick
MIT Sloan School of Management
Cambridge, MA
USA

Zeyar Aung
Masdar Institute of Science and Technology
Abu Dhabi
UAE

ISSN 0302-9743   ISSN 1611-3349   (electronic)
Lecture Notes in Artificial Intelligence
ISBN 978-3-319-13289-1   ISBN 978-3-319-13290-7   (eBook)
DOI 10.1007/978-3-319-13290-7

Library of Congress Control Number: 2014956225

LNCS Sublibrary: SL7 – Artificial Intelligence

Springer Cham Heidelberg New York Dordrecht London

Printed on acid-free paper

Springer International Publishing AG Switzerland is part of Springer Science+Business Media
(www.springer.com)

Preface

This volume contains the papers presented at DARE 2014: The Second International Workshop on Data Analytics for Renewable Energy Integration, which was held in Nancy, France in September 2014 and hosted by ECML/PKDD (the European Conference on Machine Learning and Principles and Practice of Knowledge Discovery in Databases) 2014.

Concerns about climate change, energy security, and dwindling fossil fuel reserves are stimulating ever-increasing interest in the generation, distribution, and management of renewable energy. While a lot of attention has been devoted to generation technologies, an equally important challenge is the integration of energy extracted from renewable resources into existing electricity distribution and transmission systems. Renewable energy resources like wind and solar energy are often spatially distributed and inherently variable, necessitating the use of computing techniques to predict levels of supply and demand, coordinate electricity distribution, and manage the operations of energy storage facilities.

A key element of the solution to this problem is the concept of a "smart grid." There is no standard definition but a smart grid is broadly perceived as an evolved form of the traditional electricity grid where advanced techniques, such as information and communications technology (ICT), are used extensively to detect, predict, and intelligently respond to events that may affect the supply of electricity.

Data analytics is a science that encompasses data mining, machine learning, and statistical methods, and which focuses on cleaning, transforming, modeling, and extracting actionable information from large, complex datasets. A smart grid generates a large amount of data from its various components, examples of which include renewable energy generators and smart meters; the potential value of this data is huge but exploiting this value will be almost impossible without the use of proper analytics. With the application of systematic analytics on the smart grid's data, its goal of better economy, efficiency, reliability, and security can be achieved. In other words, data analytics is an essential tool that can help to imbue the smart grid with "smartness."

In this context, the focus of DARE 2014 is to study and present the use of various data analytics techniques in the different areas of renewable energy integration. While the workshop was held on a relatively small scale it still attracted contributions that spanned a variety of relevant topical areas such as the detection of faults and other events in smart grids, forecasting of energy generation in photovoltaic and wind farms, automated analysis of rooftop PV capacity, and flexibility analysis in energy consumption. This volume will be very useful to researchers, practitioners, and other stakeholders who are seeking to leverage and drive the uptake of renewable energy and smart grid systems.

We are very grateful to the organizers of ECML/PKDD 2014 for hosting DARE 2014, the Program Committee members for their time and assistance, and to Masdar Institute of Science and Technology and MIT for their support to this timely and important workshop. Finally, we would also like to thank the authors for their valuable contributions to DARE 2014.

September 2014   Wei Lee Woon
                                                                        Zeyar Aung
                                                                     Stuart Madnick

Organization

Program Chairs

Wei Lee Woon   Masdar Institute of Science and Technology, UAE
Zeyar Aung   Masdar Institute of Science and Technology, UAE
Stuart Madnick   Massachusetts Institute of Technology, USA

Program Committee

Osman Ahmed   Siemens Building Technologies, USA
Amer Al Hinai   Sultan Qaboos University, Oman
Alvaro A. Cardenas   University of Texas at Dallas, USA
Dan Cornford   Aston University, UK
Mengling Feng   Massachusetts Institute of Technology, USA
Oliver Kramer   University of Oldenburg, Germany
Wolfgang Lehner   Technische Universität Dresden, Germany
Panos Liatsis   City University London, UK
Jeremy Lin   PJM Interconnection LLC, USA
David Lowe   Aston University, UK
Francisco Martínez-Álvarez   Pablo de Olavide University of Seville, Spain
Bruce McMillin   Missouri University of Science and Technology,
                                 USA
See-Kiong Ng   Institute for Infocomm Research, Singapore
Pierre Pinson   Technical University of Denmark, Denmark
Iyad Rahwan   Masdar Institute of Science and Technology, UAE
Kamran Shafi   University of New South Wales, Australia
Kelvin Sim   Institute for Infocomm Research, Singapore
Bogdan Trawinski   Wrocław University of Technology, Poland
Alberto Troccoli   CSIRO, Australia
Wilfred Walsh   National University of Singapore, Singapore
Paul Yoo   Khalifa University of Science, Technology
                                 and Research, UAE
Hatem Zeineldin   Masdar Institute of Science and Technology, UAE

Contents

Towards Flexibility Detection in Device-Level Energy Consumption

Bijay Neupane[✉], Torben Bach Pedersen, and Bo Thiesson

Department of Computer Science, Aalborg University, Aalborg, Denmark
{bneupane,tbp,thiesson}@cs.aau.dk

Abstract. The increasing drive towards green energy has boosted the installation of Renewable Energy Sources (RES). Increasing the share of RES in the power grid requires demand management by flexibility in the consumption. In this paper, we perform a state-of-the-art analysis on the flexibility and operation patterns of the devices in a set of real households. We propose a number of specific pre-processing steps such as operation stage segmentation, and aberrant operation duration removal to clean device level data. Further, we demonstrate various device operation properties such as hourly and daily regularities and patterns and the correlation between operating different devices. Subsequently, we show the existence of detectable time and energy flexibility in device operations. Finally, we provide various results providing a foundation for load- and flexibility-detection and -prediction at the device level.

Keywords: Device level analysis · Flexibility · Demand management

1 Introduction

Renewable energy sources (RES) are increasingly becoming an important component in the future power grids. However, with high dependence on weather conditions, such as wind and sunshine, the integration of RES into the power grid creates huge challenges, both in regards to the physical integration and in regard to management of the expected demand. Focusing on demand management, a mechanism to facilitate the energy provider with information regarding expected flexibility in users' device-usage patterns will support this management according to a fluctuating supply. On the other hand, the correlation of energy price with production from RES as in Danish Energy Market, may benefit flexible consumers by being offered lower electricity prices. Accordingly, we will in this paper analyse energy consumption at device level with the purpose of enabling flexibility detection at fine-grained resolution. We will present flexibility in device operation according to two dimensions, one dimension representing the flexibility in energy profile and a second dimension representing the flexibility in time scheduling. A flexible usage of a device will be measured according to any one or both of these dimensions, as discussed in further detailed below. Formally, we define flexibility as:

© Springer International Publishing Switzerland 2014
W.L. Woon et al. (Eds.): DARE 2014, LNAI 8817, pp. 1–16, 2014.
DOI: 10.1007/978-3-319-13290-7_1

"Flexibility: the amount of energy and the duration of time to which the device energy profile (energy flexibility) and/or activation time (time flexibility) can be changed."

Figure 1 gives an illustrative example of device level energy demand and shifting –in profile and time– to better balance fluctuating energy production from solar.

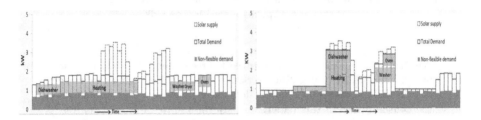

Fig. 1. Energy demand and supply, before and after demand flexibility management (using flex-offer). The non-flexible part of the demand includes lightning, TV, etc.

In order to facilitate flexibility in a dynamic energy market, the TotalFlex [7] project, implements a mechanism to express and utilize the notion of flexibility, using the concept of flex-offer [2] proposed in the EU FP7 project MIRABEL [6]. Based on a given detection or user specification of device level flexibility, this flex-offer framework addresses the challenges of balancing the variations in RES energy production and flexibility in demand by communicating a negotiation in the form of flexible consumption (or production) offers. In its simplest form, a flex-offer specifies the amount of energy required, the duration, earliest start time, latest end time, and a price. For example, "I need 2 KW to operate a dishwasher over 2 hr between 11 PM to 8 AM and I will pay 0.40 DKK/KWh for it". This flex-offer represents an example of one dimensional flexibility, which provides flexibility in activation time only but not in energy profile. The flex-offer framework further aggregates the individual flex-offers into larger units of flexibility to be traded on the reserve market. Further information on aggregation techniques and market scenarios can be found in [6,7].

The vision of the TotalFlex project is that for users having a flex-offer contract with an energy supplier, their flexibility is not specified by the user, but instead predicted within the TotalFlex architecture based on past users' behavior. The requirement for automated generation of flex-offers by detecting the flexibility in device operation is the motivation for this paper. The paper is among a very limited body of work on analyzing device level energy consumption and, to the best of our knowledge, the first with a focus on device flexibility analysis and users' device operation behaviours and patterns. More specifically, this paper focuses on the comprehensive device level analysis of energy consumption data, which will form the foundation for accurate flex-detection, flex-prediction, load-prediction, and automated generation of flex-offers. Flex-detection refers to the detection of available flexibility in device level energy consumption. Similarly, flex-prediction refers to the prediction of flexibility, e.g., an EV with a max charging power level of 5 kW is predicted to need 15 kWh of energy with a time flexibility of 8 hr starting at 09:00 PM and ending at 5:00 AM. Finally, load-prediction refers to the prediction of aggregated and device level demand for the house, e.g. the predicted energy demand for house X at 07:00 pm is 2 kWh or the predicted energy demand for oven in house X at 07:00 pm is 1.2 kWh. In summary, the main contributions of this paper are: (1) a state-of-the-art analysis of device level energy consumption, including patterns and periodicity in device operation, intra and inter device correlation, and methods to overcome data inconsistency and inadequacy in device level consumption time series, (2) a demonstration that flexible energy consumption can be detected at device level, and (3) a basis for future work of flex-detection, flex-prediction and load-prediction.

The remainder of the paper is organized as follows. Section 2 provides detailed information about existing technology and literature in device level analysis and prediction. Section 3 presents various assumptions, properties and their importance in device level analysis. In Sect. 4 we introduce the dataset on which we base our investigation. Section 5 describes the challenges associated with device level energy analysis and the preprocessing methods we have used to confront these challenges. Section 6 present various statistical analyses on device behavior. Finally, Sect. 7 concludes the paper and provides directions for future research.

2   Related Work

A review of various techniques for disaggregating household energy load was presented in [9]. Furthermore, an initiative for providing a publicly available disaggregated energy dataset has been taken in [4] together with a proposed hardware architecture for collecting household energy data. A signature based pattern matching technique to predict the device future consumption is proposed in [8]. The proposed system only predicts the power consumption for the next couple of hours for currently operating devices. In comparison, this paper performs a comprehensive device level analysis with the vision to predict the device future activation time, energy demand, and its duration of operation, for both shorter and longer time horizons (couple of days). A probabilistic model for predicting the device status, operation time, and duration is proposed in [1]. The analyses were performed with simulated data and does not address the dynamic and stochastic behavior associated with real-world device operation data. In comparison, this paper performs experiments on a real-world and freely available dataset from REDD [4]. A pool-based load shifts strategy for household appliances for leveling of fluctuating RES production has been proposed in [5]. They proposed a technique to aggregate load shift properties of similar devices into a pool known as virtual devices and use this virtual device for scheduling load shift action for appliances. In contrast, this paper focus on analyzing the patterns and periodicity associated with individual devices with a focus towards characterizing and detecting flexibility to be used in the flex-offer framework for balancing the energy market. Various methods for extracting flexibilities from electricity time series has been discussed in [3], we aim to further design flexibility detection methods at the device level.

3   Properties

We will now introduce the important device operation properties that we investigate in order to support the challenges of deriving flexibility information about energy demand. Statistical analyses for these properties will provide significant information for generalized as well as user-specific device operation patterns. Further, they provide information regarding existence of flexibility in the device operations and correlation between various devices.

1. There exists detectable Intraday and Interday patterns in device operation.
   (a) Weekend and Weekdays patterns are different.
   (b) Houses exhibit general and specific intra-day and inter-day patterns.
2. There exist time and energy flexibility in device operation.
   (a) A major percentage of energy consumption comes from flexible devices.
   (b) An alteration in device energy profile is feasible.
   (c) Device activation time can be shifted by some duration.
3. Some devices are correlated
   (a) Highly correlated device are operated simultaneously or just after one another.
   (b) There is some fixed sequence of device operation.

4   Data Description

Our analyses are performed with a disaggregated energy datasets provided by REDD [4]. We will here describe this data together with some device annotation that we have added for our experiments.

4.1   Data

The REDD dataset consists of energy consumption profiles for six different houses, each containing profiles for 16 to 24 individual devices, and is collected in April to June, 2011. Data is available for a varying number of days for the different houses; ranging from 15 to 35 days; see the third column of Table 1. Some days only have partial records, and across all houses, the total number of days with a record for at least one hour of the day is 130 days. The REDD dataset is collected at the main level, circuit level, and plug level, where the plug level data is used to log devices in cases, where multiple devices are grouped to a single circuit. The plug level data were collected using a wireless plug monitoring device and circuit level data were collected using a emonitor connected to a circuit breaker panel, as illustrated in Fig. 2. The available dataset was recorded

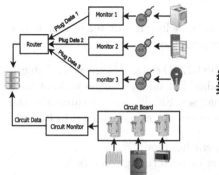 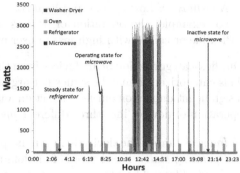

Fig. 2. The REDD hardware architecture for data collection (adapted from REDD [4]).

Fig. 3. Power demand for selected devices over the course of a day (April 30, 2011; *house 1*).

at various frequencies: 15 kHz for main phase, 0.5 Hz for circuit level, and 1 Hz for plug-level. For the main phase, data were written to the log in buffers of one second and for the circuit and plug level once every three seconds. Figure 3 shows an example of power demand for devices over the course of a day.

4.2 Device Categorization

Table 1 shows characteristics for the collected data across the individual houses in the dataset. The *Days Span* column represents the total number of days between the start and ending date in the time series, whereas *#days* is the total number of days with at least one hour of available data. Similarly, *#channels* represents the number of data collection points (plug and circuit) in a house and *#devices* is the number of unique devices available in the house.

   Our analyses of the devices in a household pertain to the possibility extracting the operational flexibility, which is evaluated based on the cost and benefit of utilizing it under the TotalFlex scenario. In that respect, we define *cost* and *benefit* as follows:

- *Cost:* The loss of user-perceived quality caused by accepting flexibility.
- *Benefit:* The available time and energy flexibility for the device.

According to this cost and benefit trade-off for flexibility, we have classified type of devices into three different *flex-categories*.

- *Fully-flexible:* High benefit at low cost
  For example, a refrigerator exhibit repeating patterns in energy consumption, which allows higher flexibility in its operation without any loss in user experience (temperature).
- *Semi-flexible:* Benefit and cost are comparable.
  For example, the flexibility in shifting activation time for the oven is associated with a cost of users' willingness to delay their cooking.

– *Non-flexible:* Low benefit and high cost
  For example, the operation of devices such as lighting or television, is not
  flexible or comes with high loss in user experience.

Our flex-categorization of all devices from the REDD dataset is shown in Table 2.
This categorization may be somewhat arguable, but minor changes will not have
a significant impact on our flex-detection analyses. Finally, we also categorize the
operation of devices into three different *operation-states* as rendered in Fig. 3:

– *Inactive State:* The device is in a non-operating mode.
– *Operating state:* The device is functioning or performing some task.
– *Steady State:* The ideal or low power consumption state between two peak
  consumption of a single operation.

Table 1. Data details for each house.

House number	Days span	#Days	#Channels	#Devices
House 1	36	35	18	11
House 2	34	15	9	9
House 3	44	23	20	13
House 4	48	30	18	12
House 5	44	9	24	15
House 6	23	18	15	11

Table 2. Device flex-categorization.

Fully-flexible	Semi-flexible	Non-flexible
Dishwasher	Furnace	Bathroom_gfi
Electric_heat	Microwave	Miscellaneous
Refrigerator	Stove	Electronics
Washer_dryer	Oven	Kitchen_outlets
		Lighting

5   Preprocessing

In this section, we will discuss the preprocessing steps and challenges associated
with the device level analysis. The complete sequence of the steps taken during the preprocessing of raw input data before allowing the statistical analysis
is visualized in Fig. 4, and details for each step are described in the following
subsections:

5.1   Spike Removal

We define noise as both being the effect of unwanted artifacts and white noise
influencing the time series data in a way that generates abnormal patterns in the
recorded device operation. Noise may be introduced through different sources,
e.g., error during data collection, abnormal behavior of the device, or mistakes
by the users (mistakenly switching device on and off). We have considered a very
high consumption value for very short duration, up to 2 data points (corresponding to 6 sec), as a noise spike. The first preprocessing step replaces the spikes in
energy consumption with its preceding neighboring value, which eliminates the

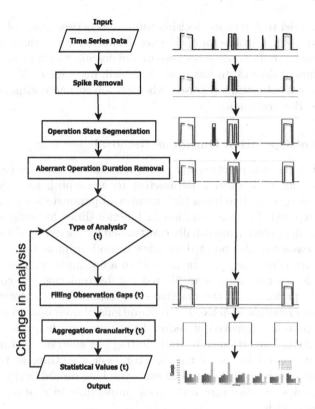

Fig. 4. Preprocessing steps.

obvious artifacts created by the installed measurement devices or user mistakes in device operation. The data, in general, contains few noise spikes of this type. To give an impression only 17 consumption spikes were removed for the oven device-data in *house 1*, which accounts for 0.004 % of total high consumption values for this device. However, removing these noise spikes significantly effects the statistical analysis, as each removed spike would otherwise be considered as an individual device operation, escalating the device operation frequency.

5.2 Operation State Segmentation

Before proceeding with device level statistical analysis, we annotate the data according to the three different states of device operation, as described in Sect. 4.2. The presence of an unknown, and for some devices, a high number of intermediate power levels between the minimum and maximum power consumption makes it challenging to segment the data into their respective phases of operation. Furthermore, variability in the power consumption value within a given stage of device operation creates ambiguity in defining the correct stage. We rely on a simple segmentation approach, where we manually inspected the consumption time series for a device in order to determine decision thresholds between the different device operation states, e.g., 50 watts for microwave, 20 watts for electric heat, etc. This manual approach is only possible because we are dealing with a reasonably small number of different devices. In future work, we will investigate the possibility of replacing this manual setting of thresholds with robust and optimal automated threshold detection techniques.

5.3 Aberrant Operation Durations Removal

Each device has its own functionality and, furthermore, the user can typically operate the device with varying parameters to accomplish an objective. For example, a cooking objective has a high number of parameters according to the meal being prepared. Despite variation in functionalities for different devices, their durations in operation are usually constrained by the device objectives, e.g., usually a dishwasher is not operated for a few seconds or more than a few hours, and a microwave is not operated for more than a few minutes. These abnormal behaviors in device operation were filtered out by replacing the consumption values in these periods with the inactive state values. In an extreme case, we discovered four instances in *house 1* with durations of oven operation that were less than 3 min. When removed, it reduced the operation frequency of oven in the house by 20 %. The rationale behind filtering out aberrant operation is that these are often caused by user's mistake, e.g. user forgetting to turn off the oven when the food is cooked or mistakenly switching on the device, and these kinds of behaviors are very rare and almost impossible to capture in analysis and prediction models.

5.4 Filling Observation Gaps

Observation gaps in the time series is a major challenge that we face for analyzing the dataset used in this paper. We have implemented three simple approaches for handling the missing information in observation gaps. The approaches differ according to the amount of missing data in the gaps. In the first case, if an observation gap stretches over a full day, that day is discarded from the analyses. Second, for observation gaps up to at most 6 sec (2 consecutive data points), this data is filled-in by simple extrapolation according to the last observed value. Finally, for observation gaps of more than 6 sec and up to a day, the imputation for the gap is based on a computation of the expected value given time-matching observed values across all days in the analysed data of the given device (and house). For gaps less than an hour, the observed data values are averaged across the hour, and for gaps greater than an hour the observed data values are averaged over a block of time for each of the five distinct periods that are shown in Table 3. In this way, we fill in $\approx 750(1.67\%)$ missing hours out of the total of $\approx 45,000$ hours for the entire dataset.

To illustrate our data imputation methods, let us first assume that we are interested in the operation state for a device and discover a gap of 20 min in a particular hour. Each of the three operation states (see Sect. 4.2) are then for the

Table 3. Division of 24 h into different block periods. The last column shows the expected value (pobability) of operation for the stove in *house 1*.

Periods	Hours of the day	Category	Probability of operation
Night	23.00 – 05.00	Off-peak period	0.03
Morning	06:00 – 08:00	On-peak period	0.67
Noon	09:00 – 13:00	Mid-peak period	0.28
Afternoon	14:00 – 17:00	Low-peak period	0.12
Evening	18:00 – 22:00	Mid-peak period	0.32

unobserved time ticks filled-in with the average observation value for each state, as computed for the considered hour across days with full observations on that hour. In another example, say, that we are still interested in operation states, but now the observation gap is a three hour period from 7:00 to 09:59. The imputation of the operation states will now proceed as in the previous example with the minor difference that the imputed values are averaged across the hours for a block period (see Table 3). Notice that in this way, the imputed values for the first two hours may be different from the imputed values for the last hour, because they belong to different block periods. Finally, to illustrate how imputed values may be involved when computing statistics in an analysis, let us consider the oven, which has imputation values for the 'operating' operation state, as shown in the last column of Table 3. Say that we observe that the oven is turned on once in the evening and we have the same three hour observation gap as above. In this case, the expected frequency for the usage of the oven during that day would be $1 + 0.67 + 0.28 = 1.79$ operations. Notice that to simplify the illustration, we have here assumed that the oven can only be turned on once during a block period and that operations do not cross blocks.

5.5 Aggregation Granularity

Finally, we aggregate the high frequency data in to the granularity of time that we target for analyses, e.g. hourly or daily. An example for aggregation of time series data into hourly resolution is shown in Fig. 5.

6 Data Analysis

In this section, we will discuss statistical behaviors regarding energy profile, usage pattern, and correlations in the device operations.

6.1 Device Energy Profiles

In Fig. 6, we show a disaggregation of the total energy consumption for the different individual devices that are measured in the REDD dataset. It shows a

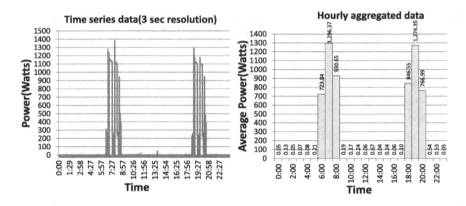

Fig. 5. Hourly aggregation of time series data for *dishwasher*.

significant variation in device consumption across the different houses. However, by aggregating according to the type of device flexibility (Fully/Semi/Non, see Table 2), as shown in Fig. 8, we see a more consistent pattern in the shares of the total energy consumed by flexible and non-flexible devices. The figure suggests that on average 50 % of the total energy demand from a house can be considered flexible, supporting Property 2(a). The low percentage of consumption from flexible, energy-intensive devices such as heating, contributed to the seasonal behavior in the device operation (late Spring to early Summer).

Figure 7 shows the minimum, maximum, and average power during operation for different devices in *house 1*. We see many devices with average consumption far from the extremes, which for all these devices indicate a potential for device designs that support altering of the energy profile during operation and, hence, creating flexibility and supporting Property 2(b). Looking at the potential for flexibility due to users' behavioral changes, additional experiments (not reported in detail here) have shown a high deviation in total energy consumption, due to duration of operation and power level, across operations for an individual device. Also, as we will see in Sect. 6.2, there can be significant variation in consumption at the time an operation is initiated. These findings indicate a flexibility potential of shifting the energy profile or activation time for the device, if users are willing to –or compensated for–behavioral changes supporting Property 2(c).

6.2   Use Patterns

We will now look more into the patterns and (ir)regularities that device operations exhibit in individual households. If we first consider the interday distribution of total energy comsumption for the individual houses' in Fig. 9, we see a general pattern of a somewhat evenly distributed consumption across the week, with a slight tendency towards lower consumption during the weekend. Exceptions are *house 5* where most of the consumption occurred on Monday and Tuesday, and *house 1* with higher consumption during weekends.

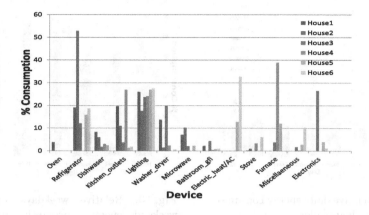

Fig. 6. Distribution of energy demand over various devices (and house).

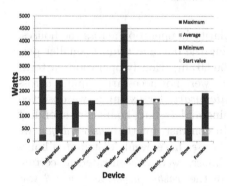

Fig. 7. Min, Avg, and Max power consumption (watts) and power consumption during the device activation (shown by dot) for selected devices

Fig. 8. Percentage distribution of total energy consumption by flexibility type from individual houses.

The weekday versus weekend difference is further emphasized by the corresponding aggregation, as shown in Fig. 10. For all but *house 1* we clearly see a higher energy consumption during the weekdays than in the weekends. This difference supports Property 1(a), but the pattern is, however, surprising in that it contrasts the common belief that people use most of their high energy consuming devices, such as washer and dryer, during weekends and holidays. It should be noted that the REDD data has a high rate of missing data during weekends and we are therefore collecting additional data to support this surprising preliminary result.

To analyze intra-day variation for individual devices within and across households, we aggregated device operations by the hours of the day in order to compute the percentage of days that a device has been in operation during a particular hour. Figure 11 shows the results obtained for the *dishwasher* operation

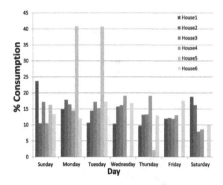

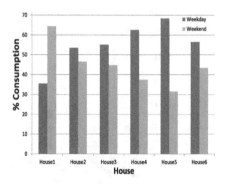

Fig. 9. Relative daily energy consumption across households.

Fig. 10. Relative weekdays versus weekends energy consumption across households.

in the six houses. We can clearly see a similarity in operation across the houses, with a higher percentage of operation during the day and a much smaller percentage during the late night (00.00 am – 06.00 am). Hourly operation patterns were observed for most devices, but patterns were not necessarily similar across different households. For the *stove* and *microwave*, for example, we observed similar peak operation periods in the morning and evening hours corresponding to the typical hours for preparing meals. In contrast, for the *washer dryer*, we found varying patterns in operation. Some houses have evenly distributed hours of operation, whereas operations in other houses were concentrated at certain hours of the day. For example, operation of the *washer dryer* in *house 3* was highly concentrated in the hours between 5.00 pm – 11.00 pm, whereas a fairly even distribution of operation was observed in *house 1*. In some cases, these peak periods represent very high percentages of device operation, as e.g., for *dishwasher* usage in *houses 2,3,5* seen in Fig. 11.

Results (not reported in detail here) also verify obvious variation in the typical duration for operation of different devices, reflecting their objectives and functionalities. For example, we see shorter operation durations for the microwave and high (continuous) duration for the refrigerator as justified by their different operational objectives. Looking at the aggregated daily frequency[1] of device operations, conclusions are household dependent. For some households we see a fairly even distribution across days, whereas for other households the frequency varies dramatically. For example, Fig. 12 shows the daily frequency of *microwave* operations for different households (*houses 4,6* do not operate this device). We see a somewhat stable usage pattern for *houses 3,5*, whereas *houses 1,2* have large daily variations.

Summarizing the intra-day results from this section, we can support Property 1(b), that device dependent intra-day patterns do exist and repeat across days. On the more challenging side, however, these patterns seems to be highly

---

[1] More precisely, the *expected* frequency, as described in Sect. 5.4.

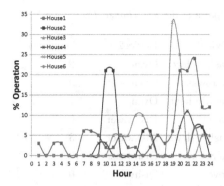

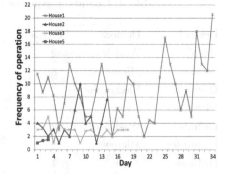

Fig. 11. Distribution of hourly *dish-washer* operations – averaged across all days in households.

Fig. 12. Daily operation frequency for *microwave* across households.

household dependent. In addition, small peak periods for operation, as we see in Fig. 11, may suggest a potential for shifting activation time, in further support of Property 2(c).

6.3 Device Correlations

Finally, we aim to provide information regarding the coherence and correlation between device operation occurring due to the usage patterns. We will further explore the sequence in which devices are operated, and there frequencies. Table 4 shows the number of times a given pair of devices were operated together within a one hour interval in *house 1*. The ordering of the pairs represents the sequence of activation of the devices. The high degree of correlation between devices such as the *stove* and *microwave* occur due to them supporting a joint activity, cooking. However, that devices such as the *washer dryer* and *microwave* exhibit a high degree of concurrent operation is more due to the user's behavior, preferences, and presence in the house. This conclusion is further justified by data from another house(not shown here), where we have very infrequent concurrent operations between *washer dryer* and *microwave*. This shows that devices can have a higher probability of concurrent operation without sharing similar functionality/purpose. Thus, we can say that occurrence of high simultaneous operation is specific to a house and does not always guarantee high correlations across households, thus providing somewhat conflicting arguments for Property 3(a). The sequential behavior in the device operations can be seen for the *microwave* and *dishwasher*, where the *microwave-dishwasher* sequence occurred 10 times in 36 days, but *dishwasher-microwave* sequence occurred only twice. This shows that the correlations in device operations vary according to their sequence of activation, supporting Property 3(b).

Table 4. Operation sequence for pairs of devices (*house 1*).

Device1	Device2	Frequency	Device1	Device2	Frequency
Oven	Washer dryer	2	Washer dryer	Oven	2
Oven	Microwave	8	Microwave	Oven	9
Oven	Electric heat	1	Electric heat	Oven	1
Dishwaser	Oven	1	Oven	Dishwaser	0
Dishwaser	Washer dryer	2	Washer dryer	Dishwaser	3
Dishwaser	Microwave	2	Microwave	Dishwaser	10
Dishwaser	Stove	1	Stove	Dishwaser	0
Washer dryer	Microwave	12	Microwave	Washer dryer	10
Washer dryer	Electric heat	1	Electric heat	Washer dryer	1
Microwave	Electric heat	8	Electric heat	Microwave	4
Microwave	Stove	6	Stove	Microwave	2
Electric heat	Stove	4	Stove	Electric heat	2
Stove	Oven	1	Oven	Stove	0

6.4 Summary

In summary, we found that a significant percentage of the total energy demand for a house can be considered to provide flexibility. Further, we noticed various repeating inter-day and intra-day, house-specific or general patterns for energy distribution and device operation across individual houses. The existence of peak operating periods for some of the devices, shows the potential of extracting time flexibility (shifting activation time) from their operation. Similarly, we observed a high variation in total energy consumption, due to varying duration of operation and power level, across device operations for an individual household. This variation in energy consumption shows the potential of extracting energy flexibility in the device operations. We also find some interesting correlations and sequences between device operation, which further provides valuable information regarding activation times of the correlated devices. Finally, we can conclude that even though there exists a stochastic behavior in device usage patterns, the patterns and periodicities can be detected and predicted, and the prediction models can be further improved by incorporating a-priori knowledge about the devices and users. The Inclusion of a-priori knowledge about the devices and users will further provide the decisive conclusion over the abnormal patterns, such as high operation frequency and abnormal energy distribution across days of the week, shown by some of the houses.

7   Conclusion and Future Work

In this paper, we investigated the flexibility in device operations with the aim of balancing fluctuating RES supply. In particular, we performed a comprehensive analysis of device flexibility and user behaviors for device operation with a focus on flex-detection, flex-prediction, and load-prediction. Flexible devices were categorized into the fully, semi, and non-flexible categories, based on the cost and benefit of extracting and using the available flexibility. Further, we proposed a number of specific preprocessing steps such as operation state segmentation, and aberrant operation duration removal, to confront the challenges associated with the analysis of device level data. For extracting the device behavior and user usage patterns, we performed three types of analyses: (1) device energy profiles were analyzed to examine the feasibility of extracting energy flexibility from household devices; (2) use patterns and periodicities were analyzed to examine the feasibility of extracting time flexibility from device operations; (3) device operation correlations were analyzed for capturing the sequences and similarities of device operation patterns. The experimental results show the general trends for houses having a significant share of consumption from flexible devices. Further, the potential of extracting time flexibility in device operations with a low loss of user-perceived quality was supported by various inter-day and intra-day energy distribution and device operation patterns, across individual houses. The experimental results support the concept of the TotalFlex project of extracting the flexibility from the device operation and utilizing it for generation of flex-offers for dynamic energy management for future energy markets.

The important directions for future work are (1) analyzing long-term data to capture seasonal behaviors in the device operations, (2) design a dynamic model for operation state segmentation to supplement the manual inspection method, (3) perform flexibility analysis for smaller resolutions, e.g. 15 min or even 1 min, (4) analyze flexibilities in RES production, and (5) perform an econometric analysis to quantify the *benefit* of utilizing flexibility, in monetary terms. Further, we consider collecting user behavior and preference data through a survey to improve the confidence over a purely data-driven flexibility analysis. Additionally, scalable and general flexibility extraction techniques and highly robust flexibility and demand prediction techniques are important to support the final goal of the Totalflex project.

Acknowledgment. This work was supported in part by the TotalFlex project sponsored by the ForskEL program of Energinet.dk.

References

1. Barbato, A., Capone, A., Rodolfi, M., Tagliaferri, D.: Forecasting the usage of household appliances through power meter sensors for demand management in the smart grid. In: SmartGridComm (2011)
2. Boehm, M., Dannecker, L., Doms, A., Dovgan, E., Filipič, B., Fischer, U., Lehner, W., Pedersen, T.B., Pitarch, Y., Šikšnys, L., Tušar, T.: Data management in the mirabel smart grid system. In: EDBT/ICDT Workshops (2012)
3. Kaulakienė, D., Šikšnys, L., Pitarch, Y.: Towards the automated extraction of flexibilities from electricity time series. In: EDBT/ICDT Workshops (2013)

4. Kolter, J.Z., Johnson, M.J.: REDD: A public data set for energy disaggregation research. In: SustKDD workshop (2011)
5. Lünsdorf, O., Sonnenschein, M.: A pooling based load shift strategy for household appliances. In: 24th International Conference on Informatics for Environmental Protection, pp. 734–743 (2010)
6. The MIRABEL Project. http://www.mirabel-project.eu
7. The TotalFlex Project. http://www.totalflex.dk/Forside/
8. Reinhardt, A., Christin, D., Kanhere, S.S.: Predicting the power consumption of electric appliances through time series pattern matching. In: BuildSys (posters) (2013)
9. Zeifman, M., Roth, K.: Nonintrusive appliance load monitoring: review and outlook. In: IEEE TCE (2011)

Balancing Energy Flexibilities Through Aggregation

Emmanouil Valsomatzis$^{(\boxtimes)}$, Katja Hose, and Torben Bach Pedersen

Department of Computer Science, Aalborg University, Aalborg, Denmark
evalsoma@cs.aau.dk

Abstract. One of the main goals of recent developments in the Smart Grid area is to increase the use of renewable energy sources. These sources are characterized by energy fluctuations that might lead to energy imbalances and congestions in the electricity grid. Exploiting inherent flexibilities, which exist in both energy production and consumption, is the key to solving these problems. Flexibilities can be expressed as flex-offers, which due to their high number need to be aggregated to reduce the complexity of energy scheduling. In this paper, we discuss balance aggregation techniques that already during aggregation aim at balancing flexibilities in production and consumption to reduce the probability of congestions and reduce the complexity of scheduling. We present results of our extensive experiments.

Keywords: Energy data management · Energy flexibility · Flex-offers · Balance aggregation

1 Introduction

The power grid is continuously transforming into a so called Smart Grid. A main characteristic of the Smart Grid is the use of information and communication technologies to improve the existing energy services of the power grid and simultaneously increase the use of renewable energy sources (RES) [2]. However, the energy generation by renewable sources, such as wind and solar, is characterized by random occurrence and thus by energy fluctuations [5]. Since their power is fed into the power grid and their increased use is a common goal, they may provoke overload of the power grid in the future, especially in peak demand situations [4,9]. Moreover, the use of new technologies, such as heat pumps and electrical vehicles (EV), and their high energy demand could also lead to electricity network congestions [13].

Within the Smart Grid, the EU FP7 project MIRABEL [3] and the ongoing Danish project TotalFlex [1] are using the flex-offer concept [17] to balance energy supply and demand. The concept is based on the idea that the energy consumption and production can not only take place in specific time slots, but be flexible and adjustable instead. For instance, an EV is parked during the night from 23:00 until 6:00. The EV could be charged or alternatively also act

© Springer International Publishing Switzerland 2014
W.L. Woon et al. (Eds.): DARE 2014, LNAI 8817, pp. 17–37, 2014.
DOI: 10.1007/978-3-319-13290-7_2

as an energy producer and feed its battery energy into the grid [15]. So the EV is automatically programmed to maximally offer 30 % of its current battery energy to the grid, corresponding to 2 h of discharging. So, in a case of an energy demand or favorable energy tariffs, the EV will be discharged, e.g. from 1:00 to 2:00, offering 15 % of its battery energy.

In the MIRABEL project, an Energy Data Management System (EDMS) is designed and prototyped. The EDMS aims at a more efficient utilization of RES by the use of the flex-offer concept. Such an EDMS is characterized by a large number of flex-offers that have to be scheduled (assign a specific time and energy amount) so that balancing supply and demand is feasible. Since it is infeasible to schedule a large number of flex-offers individually [17], an aggregation process is introduced, so that the number of flex-offers is decreased and consequently also scheduling complexity [18]. In the proposed EDMS architecture, the scheduling component is responsible for properly scheduling the aggregated flex-offers in order to balance out energy fluctuations. The TotalFlex project additionally considers balancing goal during aggregation so that imbalances are partially being handled by the balance aggregation.

The balance aggregation aims to aggregate flex-offers derived from consumption and production in order to create flex-offers with low energy demand and supply requirements. Thus, violations of the network's electricity capacities could be avoided. At the same time, the aggregated flex-offers still maintain flexibility that could further be used during scheduling to avoid grid congestions, reassure a normal grid operation, and amplify RES use. For instance, in the above mentioned EV example, there could also be another EV that needs 4 h of charging, corresponding to 70 % of its battery capacity. Charging could take place during the night from 22:00 to 5:00. So, the energy of the first EV could be used to partially recharge the second one, for example, from 23:00 to 1:00. Thus, instead of the two EVs, we consider an aggregated flex-offer that represents the demand of 2 h charging, corresponding to 40 % of battery capacity and the charge could take place from 23:00 to 5:00. In this work, we perform an extensive experimental investigation of the behavior of balance aggregation. We also propose alternative starting points for the techniques and evaluate the impact of aggregation parameters.

The remainder of this paper is structured as follows. Section 2 is describing the theoretical foundations and Sect. 3 related work. Section 4 explains how exactly balance aggregation works. Section 5 discusses results of our extensive experiments. We conclude in Sect. 6 with a discussion of our future work.

2   Related Work

The unit commitment problem [8,14], where balancing energy demand and supply is taken into consideration, has been extensively investigated through either centralized (e.g. [6,16]) or distributed approaches (e.g. [11,12]). Moreover, in [17], the unit commitment problem has been examined by handling the units as flex-offers and by using centralized metaheuristic scheduling algorithms. In [17], the economic dispatch stage of the unit commitment problem is also elaborated by applying a cost function and thus confronting potential imbalances.

Furthermore, aggregation that takes into account flexibilities with a flex-offer use case evaluation has been investigated in [19]. Scheduling aggregated flex-offers that only represent energy consumption and introducing aggregation as a pre-step of scheduling has been investigated in [18]. However, in this paper, we examine aggregation of flex-offers that takes into account one of the goals of scheduling, i.e., balancing. We do not address imbalances by using a cost function as in [18], but instead we handle imbalances as an effort to directly balance out energy amounts derived from supply and demand. To achieve that, we integrate balancing into the aggregation process. As a result, imbalances are partially handled and flexibility still remains to be used by the scheduling procedure. In this paper, we evaluate the techniques by taking into consideration energy flexibility representing not only from consumption as in [18,19] but from production as well. Moreover, this work provides an extensive experimental evaluation of balanced aggregation techniques.

3 Preliminaries

We use the following definition based on [19].

Definition 1. *A flex-offer f is a tuple $f = (T(f), profile(f))$ where $T(f)$ is the start time flexibility interval and $profile(f)$ is the data profile. Here, $T(f) = [t_{es}, t_{ls}]$ where t_{es} and t_{ls} are the earliest start time and latest start time, respectively.*

The data profile $profile(f) = s^{(1)}, \ldots, s^{(m)}$ where a slice $s^{(i)}$ is a tuple $([t_s, t_e], [a_{min}, a_{max}])$ where $[a_{min}, a_{max}]$ is a continuous range of the amount and $[t_s, t_e]$ is a time interval defining the extent of $s^{(i)}$ in the time dimension.

We consider three types of flex-offers: positive, negative, and mixed ones. Positive flex-offers have all their amount values of all their slices positive and correspond to energy consumption flex-offer. Negative flex-offers have all their amount values of all their slices negative and correspond to energy production flex-offers. All the other flex-offers are considered as mixed and express hybrid flex-offers. Figure 1 illustrates a mixed flex-offer f with four slices, $f = ([1, 7], s^{(1)}, s^{(2)}, s^{(3)}, s^{(4)})$.

The time is discretized into equal size units, e.g., 15 min and the amount dimension represents energy. Every slice is represented by a bar in the figure. The below and above bars of each slice represent the minimum and maximum amount value, a_{min} and a_{max}, respectively. A flex-offer also supports a lowest and a highest total amount that represents the minimum and the maximum energy required respectively [10].

Moreover, as defined in [19], the *time flexibility*, $tf(f)$, of a flex-offer f, is the difference between the latest and earliest start time, the *amount flexibility*, $af(s)$, is the sum of the amount flexibilities of all slices in the profile of f, and the *total flexibility* of f is the product of the time flexibility and the amount flexibility, i.e. $flex(f) = tf(f) \cdot af(s)$. For instance, the flex-offer in Fig. 1 has $tf(f) = 7 - 1 = 6$, $af(s) = (3 - 1) + (4 - 2) + (2 - (-4)) + (-1 - (-3)) = 12$, and total flexibility: $6 * 12 = 72$.

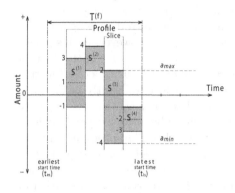

Fig. 1. A mixed flex-offer

Furthermore, we consider as aggregation the process in which the input is a set of flex-offers and the output is also a set of flex-offers (aggregated) with a smaller or equal number of flex-offers. An aggregated flex-offer encapsulates one or more flex-offers and is able to describe the flex-offers that created it. Furthermore, a measurement that we use to evaluate the aggregation is the flexibility loss that is defined according to [19] as the difference between the total flex-offer flexibility before and after aggregation.

We will further discuss the aggregation process, introduce balance aggregation, and two of its techniques in Sect. 4.

4   Balance Aggregation

In order to describe balance aggregation we define the balance, $balance(f)$, of a flex-offer f, as the sum of the average amount values of each slice, and absolute balance, $absolute\_balance(f)$ as the sum of the average absolute values of the amounts for each slice of the flex-offer. For example in Fig. 1, the flex-offer $f = ([1,6], s^{(1)}, s^{(2)}), s^{(3)}, s^{(4)})$ has $balance(f) = 1 + 3 - 1 - 2 = 1$ and $absolute\_balance(f) = |1| + |3| + |-2| + |-1| = 9$. Goal of the balance aggregation is to create aggregated flex-offers with low values of absolute balance. In this section, we sketch the aggregation process and describe approaches to achieve balance aggregation.

4.1   Flex-Offer Aggregation

In Fig. 2 we present a simple aggregation scenario where we aggregate two flex-offers, f_1 and f_2, creating the aggregated flex-offer $f_{1,2}$. Both f_1 and f_2, have time and amount and so does the aggregated one. As illustrated, the time flexibility (hatched area) of f_1 is 2 and of f_2 is 3. In the first column of the figure we show the *start alignment* aggregation [19]. In that case, we align the two flex-offers so that their profiles start at the earliest start time and then we sum the minimum and maximum amounts of each aligned slice. The time flexibility of the aggregated flex-offer is the minimum flexibility of the non-aggregated flex-offers. This reassures that all the possible positions of the *earliest start time* of the aggregated flex-offer will not violate the time constraint that the non-aggregated flex-offers have.

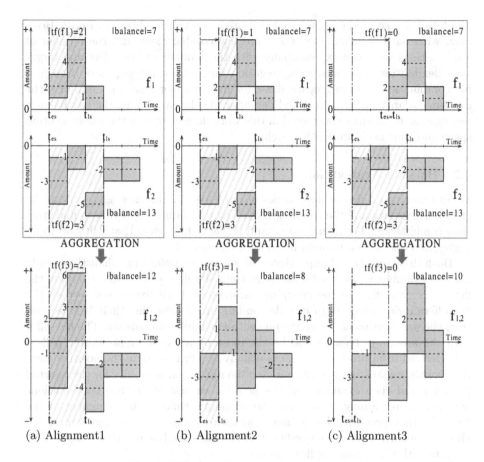

Fig. 2. Different alignments for aggregation

However, because the time flexibility of the aggregated flex-offer depends on the minimum flexibility of the flex-offers that participate in the aggregation, the flex-offers are grouped according to 2 different parameters, EST (*Earliest Start time Tolerance*) and TFT (*Time Flexibility Tolerance*) to minimize flexibility losses [19]. The value of EST represents the maximum difference of the earliest start times that the flex-offers could have in order to belong to the same group. The value of TFT represents the maximum time flexibility differences that the flex-offers could have in order to be grouped together.

As we can see in the second and the third column of Fig. 2, there are different start time profile combinations for the flex-offers that participate in the aggregation and result in different aggregated flex-offers. Note that the absolute balance of the aggregated flex-offer also depends on the start time profile combinations. For instance, in the second column where we shift the first flex-offer for one time unit, we see that the absolute balance of the aggregated flex-offer has reduced. On the other hand, continuing shifting the first flex-offer will increase again the absolute balance of the aggregated flex-offer, see third column in Fig. 2. However, with only two flex-offers there are few combinations and the number of combinations increases exponentially with the number of flex-offers and larger time flexibilities. The balance aggregation aims at identifying start time profile combinations and aggregate flex-offer in a manner that will minimize the absolute balance of the aggregated flex-offer. At the same time, the two balance aggregation techniques considered in this work do not explore the whole solution space and thus avoid in-depth search.

4.2 Balance Aggregation

We examine two different approaches of implementing balance aggregation, the exhaustive greedy and the simple greedy. The techniques are trying to find start time combinations between positive and negative flex-offers that will lead to an aggregated one with a minimum absolute balance.

Both these greedy techniques have the same start point but exhaustive greedy aims to examine a larger solution space than simple greedy. After grouping the flex-offers according to the grouping parameters, both techniques sort all the flex-offers inside each group in a descending order regarding their balance and start the aggregation choosing the one with the minimum balance. The flex-offer with the minimum balance will be the most negative one representing a flex-offer derived from production that is usually less flexible than the positive ones. Afterwards, exhaustive greedy considers the maximum flexibility for the selected flex-offer and then examines all the possible earliest start time combinations with all the remaining flex-offers. The technique chooses the alignment of the flex-offer that provides the minimum absolute balance. It continues until the absolute balance is no longer reduced. If the absolute balance is not reduced, it restarts with the remaining flex-offers.

On the other hand, simple greedy also starts the aggregation by choosing the flex-offer with the minimum balance. However, aggregation continues with the flex-offer that has the balance which is closest to the opposite ($\pm$) balance of the first one. It examines all the possible start time combinations of the flex-offer and chooses the aggregation that has the minimum absolute balance. It also continues the aggregation until the absolute balance of the aggregated flex-offer is not further reduced. In case the absolute balance is not reduced, it considers the aggregated flex-offer and continues with the remaining flex-offers. In Fig. 3 we show how exhaustive greedy and simple greedy work in the same group of flex-offers. The example is from one of our datasets and in the figure we show the balance of each flex-offer. We see that both the techniques start their aggregation by choosing the most negative flex-offer, f_5, and aggregate it with flex-offer f_1. However in step 3, exhaustive greedy chooses to aggregate with f_3 because it gives a better absolute balance and simple greedy with flex-offer f_2 since it is closer to the opposite of the f_{51} balance. As a result, exhaustive greedy continues the aggregation separately for f_2 and f_4 and simple greedy continues aggregation with f_3 leaving f_4 non-aggregated. Therefore, the two techniques lead to different absolute balance values.

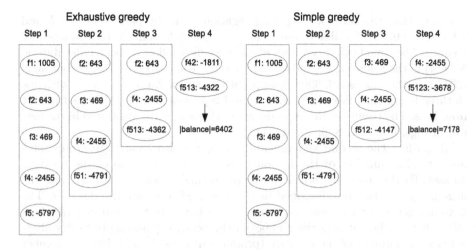

Fig. 3. Exhaustive and simple greedy examples

5 Experimental Evaluation

In this section, we present an extensive experimental evaluation of the balance aggregation techniques a comparison to start alignment aggregation discussed in Sect. 4.

5.1 Experimental Setup

For the evaluation of the balance aggregation techniques, we used 4 extensive experimental setups of 10 groups of 8 datasets, 320 datasets in total. Each setup is characterized by different time and amount probabilistic distributions corresponding to different energy scenarios.

The first experimental setup is based on the one described in [19]. It consists of 10 groups and each group has 8 datasets, 80 in total. In order to create each group of the 8 datasets, we first select flex-offers derived from the historical consumption time series of 5 random customers. Then, we incrementally add to each dataset flex-offers corresponding to the number of 5 more random customers. The last one, the eighth dataset, has flex-offers derived from historical consumption time series of 40 random customers. Afterwards, for every dataset we apply start alignment aggregation according to [19], with EST and TFT equal to zero, resulting in an aggregated positive flex-offer. In every aggregated flex-offer, a random number of amount slice, between zero and its time flexibility value is added. Finally, all the positive aggregated flex-offers are converted to negative ones. As a result, there are always one or more positive flex-offers that if being aggregated have the same opposite balance as a negative one. Since we add in each dataset flex-offers derived from five more customers, the datasets have an incremental number of flex-offers that approximately corresponds to 11 K additional flex-offers for every 5 customers. The way the dataset is created reassures that whenever we apply aggregation with the parameters EST and TFT set to 0, it is feasible to create an aggregated flex-offer with zero absolute balance. The time flexibility values of the flex-offers follow a normal distribution N (8, 4) in the range [4, 12] and the number of the slices a normal distribution N (20, 10) in the ranges [10, 30]. Those profiles are from 2.5 to 7.5 h long, with one to three hours time flexibility, which could represent flex-offers derived mostly from EVs. The results of the experiments of this setup are illustrated in the first row of Figs. 4, 5, 6, 7, 8, 9, 10, and 11.

Regarding the second experimental setup, we also created 10 groups of 8 datasets. The number of the flex-offers is similar to the one of the previous dataset. Furthermore, historical consumption time series of customers and the flex-offer generator tool described in [7] were used to create the datasets. The flex-offer generator tool was used to generate both positive and negative flex-offers. For all the datasets the number of the positive (consumption) flex-offers is twice the number of the negative (production) ones. In addition, the number of the slices of the positive flex-offers follows the normal distributions $\mathcal{N}(20, 10)$ in the ranges [10, 30] and of the negative flex-offers the normal distributions $\mathcal{N}(40, 20)$ in the ranges [20, 60]. The time flexibility values $(t_{ls} - t_{es})$ of the positive flex-offers and of the negative flex-offers follow a discrete uniform distribution on the interval [1, 10] and [1, 8] respectively. This setup aims to explore a scenario in which balancing out energy and production is theoretically feasible. Thus, the negative flex-offers are half the positive ones but with double profile length. Such negative flex-offers with long profiles and less flexibility than the flex-offers representing the consumption could simulate RES. On the other hand, flex-offers characterized by more time flexibility and shorter profiles represent flex-offers derived from mostly recent technological achievements such as EVs and heat pumps. The results of the experiments of this setup are illustrated in the second row of Figs. 4, 5, 6, 7, 8, 9, 10, and 11.

Our third experimental setup is created as the second one. These datasets are variations of the second one with a deviation regarding the length and the time. More specifically, the slices of the positive flex-offers follow the same distribution as before, but the negative flex-offers follow the normal distribution $\mathcal{N}(50, 10)$ in the ranges [40, 60], which makes them longer. The time flexibility values $(t_{ls} - t_{es})$ of the positive flex-offers and of the negative flex-offers follow a discrete uniform distribution on the interval [2, 18] and [1, 6] respectively, making the positive flex-offers much more flexible regarding time. Such kind of flex-offers could represent not only EVs and heat pumps but also electronic devices as well. The results of the experiments of this setup are illustrated in the third row of Figs. 4, 5, 6, 7, 8, 9, 10, and 11.

Our last experimental setup is similar to the third one. However, the negative flex-offers in this setup are characterized by less energy flexible profiles compared to the positive ones, reflecting a scenario in which the RES are not that flexible regarding energy. Moreover, the positive flex-offers are twice the number of the negative ones. More specifically, the number of the slices for the positive and the negative flex-offers follow the normal distributions $\mathcal{N}(5, 2)$ and $\mathcal{N}(10, 2)$ in

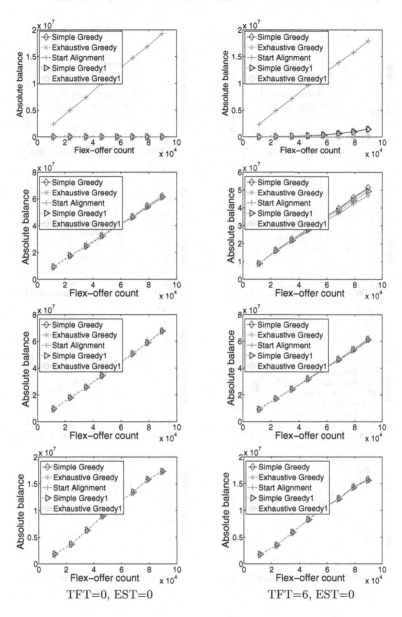

TFT=0, EST=0   TFT=6, EST=0

Fig. 4. Results of the absolute balance in terms of scalability effect

the ranges $[1, 10]$ and $[5, 15]$, respectively. The energy flexibility values of the positive flex-offers follow the normal distributions $\mathcal{N}(30, 10)$ in the range $[0, 50]$ and of the negative flex-offer $\mathcal{N}(20, 10)$ in the same range over the same amount of flexibility. The results of the experiments of this setup are illustrated in the fourth row of Figs. 4, 5, 6, 7, 8, 9, 10, and 11.

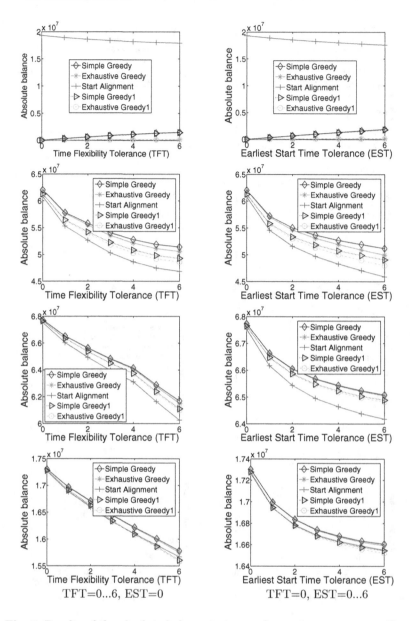

Fig. 5. Results of the absolute balance in terms of grouping parameters effect

In the experiments we investigate the three aggregation techniques regarding the absolute balance, the flexibility loss, the number of the aggregated flex-offers, and the processing time. We examine all four aspects in terms of scalability and grouping parameters, *EST*, and *TFT*. In terms of scalability, we set both the grouping parameters to zero, and examined the techniques in datasets with

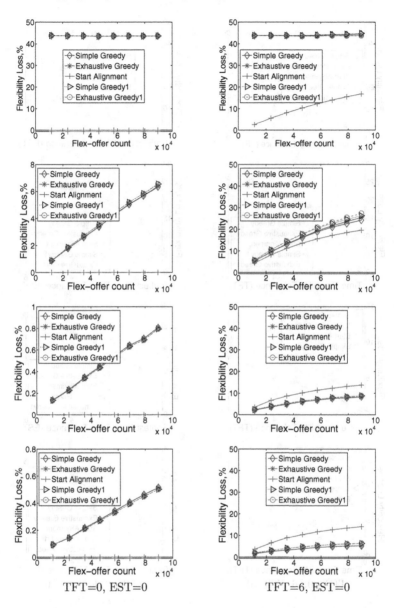

TFT=0, EST=0   TFT=6, EST=0

Fig. 6. Results of the flexibility loss in terms of scalability effect

incremental numbers of flex-offers, starting with minimum 11 K (approximately) and maximum 90 K (approximately) flex-offers (Figs. 4, 6, 8, and 10). For each experimental setup, we created 10 groups of datasets that have the same number of flex-offers to reduce any effect of the randomness that characterizes the dataset generation. Regarding the effect of the grouping parameters, we used datasets

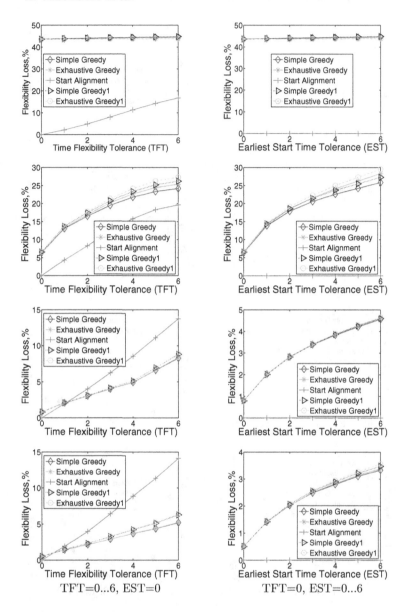

Fig. 7. Results of the flexibility loss in terms of grouping parameters effect

with almost 90 K flex-offers and set one of the parameters stable and set to zero, and varied the other values from zero to six, respectively (Figs. 5, 7, 9, and 11). For illustrating purposes, we show the average behavior of the similar datasets that there are in each group.

We also investigate the performance of exhaustive and simple greedy after alternating their starting point referring to them in all the figures as "exhaustive

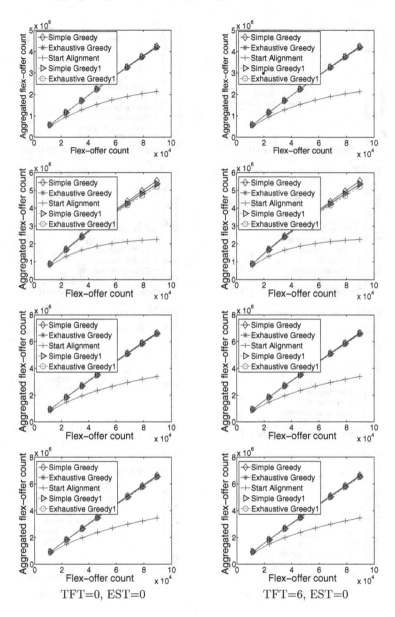

Fig. 8. Results of the aggregated flex-offer count in terms of scalability effect

greedy" and "simple greedy1". We illustrate their performance when they both start by selecting the flex-offer with the maximum absolute balance instead of the one with the minimum balance. In the first row, and the first and second column of Figs. 4, 6, and 8 there is an overlap between the illustrated lines of the techniques because the techniques showed similar behavior. The experiments

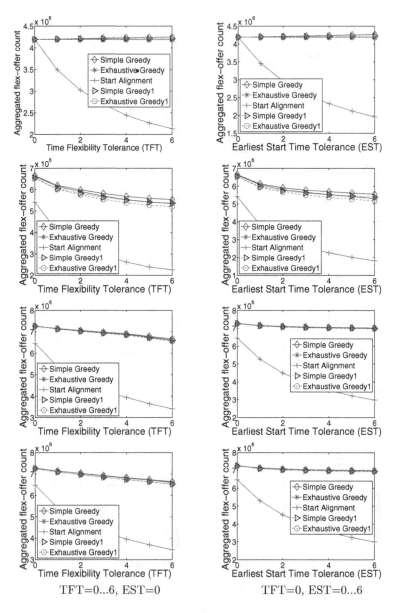

Fig. 9. Results of the aggregated flex-offer count in terms of grouping parameters effect

were conducted on a 2.9 GHz Intel core i7 processor with two cores, L2 Cache of 256 KB, L3 Cache of 4 MB and physical memory of 8 GB (4 of 4 GB of 1600 MHz DDR3).

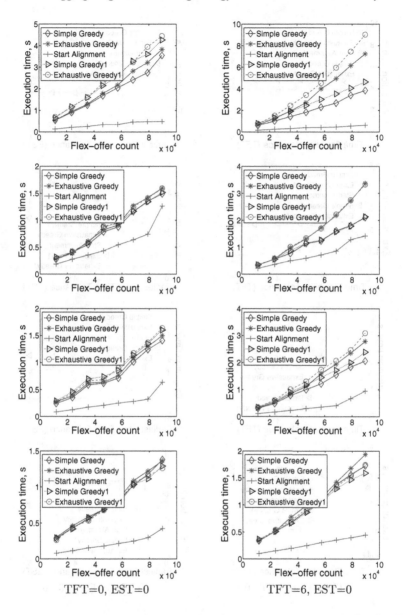

Fig. 10. Results of the processing time in terms of scalability effect

5.2 Absolute Balance

The results for absolute balance are shown in Figs. 4 and 5. In Fig. 4, absolute balance scales almost linearly with the number of input flex-offers. All the techniques, exhaustive greedy, simple greedy and start alignment have almost the same performance (first row of Fig. 4) in all the setups except the first one.

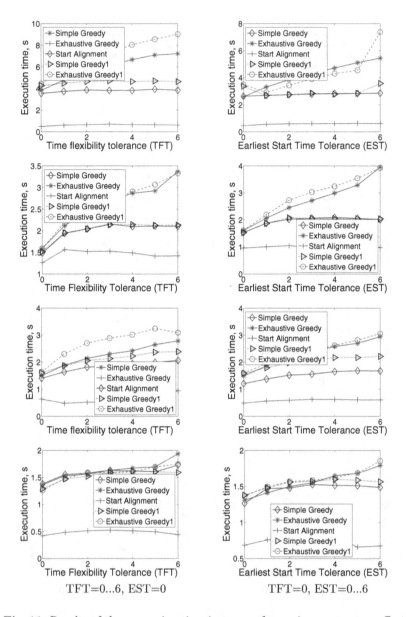

Fig. 11. Results of the processing time in terms of grouping parameters effect

In the first setup, the two greedy techniques, exhaustive and simple greedy, achieve a very low balance, first row of Figs. 4 and 5.

More specifically, we see in the first column of Fig. 4 that both techniques achieve a very close to zero balance when both EST and TFT are set to zero. When TFT is set to 6, exhaustive greedy achieves a lower balance than simple greedy (first row, second column in Fig. 4), but close to zero for both. Due to the nature of the first experimental setup, zero absolute balance can be achieved and therefore the two greedy techniques achieve it. However, we observe that start alignment achieves the minimum absolute balance among the techniques in the last three setups when there is a large number of flex-offers, approximately 90 K, see Fig. 5. In the second to fourth line of Fig. 5 we see that exhaustive greedy and simple greedy have absolute balance similar to start alignment, with exhaustive greedy achieving a slightly smaller value between the two greedy techniques.

In the last three experimental setups (second to fourth row in Figs. 4 and 5), there is an overlap in all the slices between the positive and the negative flex-offers since the profiles of the negative flex-offers have at least double profile length in comparison to the positive ones. As a result, start alignment achieves a low absolute balance and the two greedy techniques do not take advantage of the exploration of the solution space since all the differences in absolute balance between each possible aggregation are very low. They are very low because even if there might be an earliest start time combination between a positive and a negative flex-offer that reduces the total absolute balance of the mixed flex-offer, the percentage will be too low because the absolute balance is mostly reflected by the long profiles of the negative flex-offer. This fact combined with the large number of aggregated flex-offers that the two greedy techniques produce after aggregation in all the datasets, (Figs. 8 and 9) produce a lower balance for start alignment.

The number of the flex-offers that is produced by aggregation influences the result of the absolute balance. The fewer flex-offers participate in the aggregation, the more aggregated are produced. As a result less compensations between positive and negative slices take place and that leads to a higher absolute balance. Furthermore, start alignment shows a decreasing behavior of the produced absolute balance when the values of the grouping parameters are greater than zero, first row of Fig. 9. This occurs because when the values of the grouping parameters increase, more flex-offers participate in the aggregation and thus more positive and negative flex-offers are aggregated, achieving a lower absolute balance.

5.3 Flexibility Loss

Another aspect that we examine for the aggregation is the flexibility loss. In Figs. 6 and 7 we see that the techniques show a divergent behavior in all the different setups. We see, in the first column of Fig. 6, that start alignment has zero absolute balance when $TFT = 0$, followed by exhaustive greedy that has a small difference to simple greedy. The flexibility loss is mainly affected by the time flexibility and since both EST and TFT are set to zero, it means that in each group the flex-offers have the exact same earliest start time and the same time flexibility as well. That leads to no time flexibility loss for start alignment and thus no flexibility loss at all. Furthermore, we notice a low percentage of flexibility loss for exhaustive and simple greedy in the three last setups (second column, second and third row of Fig. 6 and first column, second and third row of Fig. 7). The low time flexibility of the negative flex-offers of the third and the fourth setups reassures a low flexibility loss. This happens because the solution space is narrowed down, low time flexibility leads to fewer combinations, and that TFT set to 0 reassures that all the flex-offers in the group have the same low time flexibility. A higher percentage of flexibility loss for both greedy techniques is shown in the second setup (second row of Figs. 6 and 7), because in this dataset the time flexibility of the flex-offers is higher.

Both exhaustive and simple greedy behave similarly to start alignment, producing almost the same number of aggregated flex-offers (first column of Fig. 8). However, we notice a high percentage of the flexibility loss for both greedy and simple greedy in the first setup (first row of Figs. 6 and 7). We see that start alignment has, as before, zero flexibility loss, and the nature of the setup favors an exploration of the solution space for both greedy techniques. Eventually, the greedy techniques identify aggregations that lead to less time flexibility and thus to flexibility loss.

Based on the second column of Fig. 6, the first and partially the second column of Fig. 7, we notice that for all the techniques, when the grouping parameters are increased, the flexibility loss is also increased. This happens because flex-offers with different time flexibilities are in the same group. For start alignment, this will lead to an aggregated flex-offer with the lowest time flexibility and hence to flexibility loss. Regarding exhaustive and simple greedy, larger grouping parameters result in a larger solution space since more flex-offers participate in the aggregation and more earliest start time combinations exist. As a result, the techniques will most probably create an aggregated flex-offer with a lowest absolute balance and a lowest time flexibility. However, no matter how much we increase the value of the EST parameter, the fact that TFT is zero will reassure the maintenance of the time flexibility for start alignment and thus no flexibility loss will occur. In almost all the datasets, start alignment shows the best behavior compared to the other two techniques. In the third and the fourth experimental setup, we see that while the number of the flex-offers increases and especially while TFT increases (third and fourth row of Fig. 7), the two greedy techniques show a result that is competitive to start alignment result and even better, achieving a lower flexibility loss. The low flexibility losses occur due to the high value of the time flexibility that the positive flex-offers are characterized with, compared to the negative ones in the third and the fourth setup. Therefore, there are flexibility losses for start alignment, since the aggregated flex-offers have the lowest time flexibility.

The two greedy techniques achieve a lower flexibility loss because some of the flex-offers in the group do not participate in the grouping. Hence, exhaustive and simple greedy create more aggregated flex-offers than start alignment (Figs. 8 and 9), thus fewer flex-offers participate in the aggregation and less flexibility loss will occur.

5.4   Execution Time and Aggregated Flex-Offers Count

Regarding the processing time of all the techniques, we see in Figs. 10 and 11 that start alignment has the best performance followed by simple greedy and exhaustive greedy. Start alignment is the fastest one since it always applies only one aggregation. It is also possible for start alignment to achieve better execution times, see third row, second column of Fig. 11, when the grouping parameters are high and thus fewer groups are created. On the other hand, exhaustive greedy demands the most execution time because it creates more than one aggregated flex-offers as simple greedy does, but explores a larger solution space than simple greedy. This results to larger execution times for exhaustive greedy, leaving simple greedy in the second place. Regarding the number of the aggregated flex-offers, we see in Figs. 8 and 9 that in all the experimental setups, start alignment has a lower number of aggregated flex-offers than the two greedy techniques. This is a result of the implementation of the techniques because start alignment will always create one aggregated flex-offer when it is applied to a set of a flex-offers. On the other hand, exhaustive and simple greedy will create at least one aggregated flex-offer if absolute balance is not reduced during the aggregation. Regarding the alternative exhaustive and simple greedy we see no difference for the first experimental setup. However, in all the other setups, the alternate techniques achieve a better absolute balance, higher flexibility losses, and fewer aggregated flex-offers when the grouping parameters are set to values greater than zero for larger number of flex-offers (second to fourth row of Figs. 5, 7, and 9). On the other hand, since they create fewer aggregated flex-offers, they have a bigger solution space to examine and thus they are slower than the original ones (Figs. 10 and 11).

6   Conclusion and Future Work

In this paper, we elaborated on aggregation techniques that take into account balancing issues. The techniques discussed in Sect. 4 reduce the number of the flex-offers that will be the input of the scheduling process and at the same time consider one of its main goals, i.e., achieving balance between energy supply and demand. We conclude through an extensive experimental evaluation that achieving the minimum balance is feasible, but there is always a trade off between balance, flexibility loss and processing time. We show that in order to achieve a good balance, we have to sacrifice time flexibility and also spend more time on processing. The comparisons of the balance techniques with start alignment aggregation showed as well that there are scenarios in which start alignment can achieve very good balance in faster processing times than the greedy techniques. However, flexibility loss between the techniques depends on the grouping parameters without providing a clear winner.

In our future work, we aim to improve the grouping phase that takes place in order to maximize the flexibility that the aggregated flex-offers will have and at the same time improve the balance. It seems also interesting to examine the balance that aggregation will achieve during hierarchical aggregation that is important for an EDMS. In such a scenario, balance aggregation seems more suitable since the input will be mixed flex-offers.

Acknowledgments. This work was supported in part by the TotalFlex project sponsored by the ForskEL program of Energinet.dk.

References

1. Totalflex project. http://www.totalflex.dk/
2. Bach, B., Wilhelmer, D., Palensky, P.: Smart buildings, smart cities and governing innovation in the new millennium. In: 8th IEEE International Conference on Industrial Informatics (INDIN), pp. 8–14 (2010)
3. Boehm, M., Dannecker, L., Doms, A., Dovgan, E., Filipic, B., Fischer, U., Lehner, W., Pedersen, T.B., Pitarch, Y., Šikšnys, L., Tušar, T.: Data management in the mirabel smart grid system. In: Proceedings of EnDM (2012)
4. European Wind Energy Association: Creating the internal energy market in Europe. Technical report (2012). http://www.ewea.org/uploads/tx_err/Internal_energy_market.pdf
5. Hermanns, H., Wiechmann, H.: Future design challenges for electric energy supply. In: IEEE Conference on Emerging Technologies Factory Automation, pp. 1–8 (2009)
6. Hosseini, S., Khodaei, A., Aminifar, F.: A novel straightforward unit commitment method for large-scale power systems. IEEE Trans. Power Syst. **22**(4), 2134–2143 (2007)
7. Kaulakienė, D., Šikšnys, L., Pitarch, Y.: Towards the automated extraction of flexibilities from electricity time series. In: Proceedings of the Joint EDBT/ICDT 2013 Workshops, pp. 267–272. ACM (2013)
8. Kazarlis, S., Bakirtzis, A., Petridis, V.: A genetic algorithm solution to the unit commitment problem. IEEE Trans. Power Syst. **11**(1), 83–92 (1996)
9. Kupzog, F., Roesener, C.: A closer look on load management. In: 5th IEEE International Conference on Industrial Informatics, vol. 2, pp. 1151–1156 (2007)
10. Siksnys, L., Thomsen, C., Pedersen, T.B.: MIRABEL DW: managing complex energy data in a smart grid. In: Cuzzocrea, A., Dayal, U. (eds.) DaWaK 2012. LNCS, vol. 7448, pp. 443–457. Springer, Heidelberg (2012)
11. Logenthiran, T., Srinivasan, D., Khambadkone, A., Aung, H.N.: Multiagent system for real-time operation of a microgrid in real-time digital simulator. IEEE Trans. Smart Grid **3**(2), 925–933 (2012)
12. Logenthiran, T., Srinivasan, D., Khambadkone, A.M.: Multi-agent system for energy resource scheduling of integrated microgrids in a distributed system. Electr. Power Syst. Res. **81**(1), 138–148 (2011)
13. Lopes, J., Soares, F., Almeida, P.: Integration of electric vehicles in the electric power system. Proc. IEEE **99**(1), 168–183 (2011)
14. Padhy, N.: Unit commitment-a bibliographical survey. IEEE Trans. Power Syst. **19**(2), 1196–1205 (2004)
15. Rezaee, S., Farjah, E., Khorramdel, B.: Probabilistic analysis of plug-in electric vehicles impact on electrical grid through homes and parking lots. IEEE Trans. Sustain. Energ. **4**(4), 1024–1033 (2013)
16. Srinivasan, D., Chazelas, J.: A priority list-based evolutionary algorithm to solve large scale unit commitment problem. In: International Conference on Power System Technology, vol. 2, pp. 1746–1751 (2004)

17. Tušar, T., Dovgan, E., Filipic, B.: Evolutionary scheduling of flexible offers for balancing electricity supply and demand. In: 2012 IEEE Congress on Evolutionary Computation (CEC), pp. 1–8 (2012)
18. Tušar, T., Šikšnys, L., Pedersen, T.B., Dovgan, E., Filipič, B.: Using aggregation to improve the scheduling of flexible energy offers. In: International Conference on Bioinspired Optimization Methods and their Applications, pp. 347–358 (2012)
19. Šikšnys, L., Khalefa, M.E., Pedersen, T.B.: Aggregating and disaggregating flexibility objects. In: Ailamaki, A., Bowers, S. (eds.) SSDBM 2012. LNCS, vol. 7338, pp. 379–396. Springer, Heidelberg (2012)

Machine Learning Prediction of Large Area Photovoltaic Energy Production

Ángela Fernández[✉], Yvonne Gala, and José R. Dorronsoro

Dpto. Ing. Informática, Universidad Autónoma de Madrid, Madrid, Spain
a.fernandez@uam.es

Abstract. In this work we first explore the use of Support Vector Regression to forecast day-ahead daily and 3-hourly aggregated photovoltaic (PV) energy production on Spain using as inputs Numerical Weather Prediction forecasts of global horizontal radiation and total cloud cover. We then introduce an empirical "clear sky" PV energy curve that we use to disaggregate these predictions into hourly day-ahead PV forecasts. Finally, we use Ridge Regression to refine these day-ahead forecasts to obtain same-day hourly PV production updates that for a given hour h use PV energy readings up to that hour to derive updated PV forecasts for hours $h + 1, h + 2, \ldots$. While simple from a Machine Learning point of view, these methods yield encouraging first results and also suggest ways to further improve them.

Keywords: Photovoltaic energy · Numerical weather prediction · Support vector regression · Ridge regression

1 Introduction

The constant integration of renewable energy, particularly wind and solar, and their increasing effect in the electrical systems of countries such as the USA, Germany, Denmark or Spain has as a consequence a growing needed of accurate forecasts. Usually these forecasts are needed for the day-ahead level, i.e., about 24 h in advance, or for same-day hourly updates of previous forecasts that are given for the next few hours. In the case of Spain, 24 h day-ahead forecast are required for daily energy markets and same-day 4–8 h updates for intra-day markets. A large forecast effort has been carried out in the past years for wind energy and has resulted in a wide use of Machine Learning (ML) models and tools such as Neural Networks (NN) or Support Vector Regression (SVR) to predict either local energy production at single wind farms or global energy production over a wide geographic area. These NN or SVR models have as their inputs the day ahead Numerical Weather Prediction (NWP) forecasts provided by systems such as the Global Forecasting System (GFS, [5]) or the European Center for Medium Weather Forecast (ECMWF, [4]) and give very good results for day ahead prediction. On the other hand, short term wind energy forecasting is more difficult, as simple persistence prediction is very hard to beat at the first

© Springer International Publishing Switzerland 2014
W.L. Woon et al. (Eds.): DARE 2014, LNAI 8817, pp. 38–53, 2014.
DOI: 10.1007/978-3-319-13290-7_3

1–2 h, simple models are usually competitive for the next few hours and any model reverts to the NWP-based day-ahead forecast in about 5–8 h at single wind farms and 10–14 h for very wide area predictions. We point out that wind farms are usually not operated, in the sense that wind energy is directly converted into electric energy without the farm's output being regulated. Thus, the farm works on a stable basis and model building and prediction can thus be carried through in a time homogeneous basis.

Solar energy forecasting is following basically the steps already taken for wind energy. There are two main solar technologies, thermosolar or concentrating solar power (CSP), where the Sun's radiation is used to heat up a fluid that then drives a steam turbine to produce electricity, and photovoltaic (PV), where radiation is transformed directly into electricity. CSP plants are usually operated to a high degree, particularly if they have facilities such as molten salt tanks to store heat that can be used to generate energy after dark. This makes obviously quite difficult their direct energy forecasting by, say, ML methods. On the other hand, PV plants are at this moment not so operable and, thus, are in principle more amenable to direct ML energy forecasts. However, they are relatively small and, thus, very sensitive to cloud effects. Moreover, PV energy production has no inertia, which results in very sharp and wide fluctuations which as of today, are essentially impossible to predict for single plants. A further difficulty is the fact that solar NWP forecasting is much less developed than wind NWP; in fact, wind energy has been a driving force in the past few years for the NWP systems to adapt to the industry needs. However, this is not yet the case with radiation, cloud or aerosol modeling.

In any case, the efficiency of PV cells is constantly improving and the costs of installation and operation of roof top micro PV plants are falling. This will likely lead to a potentially large increase of small and decentralized but still grid-connected micro plants. To some extent this is already the case in Spain, where there are about 4 GW of installed PV power but perhaps less than 1 GW corresponds to "large" (i.e., above 2 MW) PV plants. The local prediction of their output will be probably a too challenging problem in the near future; on the other hand, the accurate forecasting of the PV output of a much wider area should be more manageable.

In this paper we consider a particular and very large case, the hourly prediction of the global PV energy over peninsular Spain. We will consider both the day-ahead scenario, where energy predictions for the 24 h of day d are given some time at day $d-1$, and the same-day (or short-term) prediction updates at a given hour h of previous day-ahead forecasts for hours $h+1, h+2, \ldots$. We shall use as model inputs in this first case the ECMWF NWP predictions of two meteorological variables, the aggregated downward surface solar radiation (DSSR) and the average total cloud cover (TCC), given every three UTC hours (i.e., 8 values per day) over the 1,128 points of a grid that covers essentially the entirety of the Iberian peninsula. More precisely, the DSSR values at hour h contain the sum of radiation at hours $h-2, h-1, h$; a three hour TCC value gives the sky fraction covered at that time by clouds.

There is a growing number of publications on statistical and ML modeling of radiation (certainly the most relevant variable for PV energy) but also of PV energy itself. Two recent examples of such a use of ML methods are the benchmark performed under the Weather Intelligence for Renewable Energies WIRE COST Action[1], and the 2013 competition jointly organized by the American Meteorological Society and the company Kaggle[2]. In the first case, Quantile Regression was used to derive PV estimates under a clear sky assumption. In the second, the goal was to predict daily aggregated radiation at a number of weather stations in Oklahoma from an ensemble set of NWP predictions; the winning model had as its basis Gradient Boosting Regression. On the other hand, NNs are applied in [1] to predict daily average radiation values. With respect to PV prediction, [9] reviews several Artificial Intelligence techniques in PV energy and [11,12] consider a number of ML methods. A good general reference for the many issues present in radiation and PV energy is the recent book [7]. We also point out to the 2013 Data Analytics for Renewable Energy workshop, where [14] gives a thorough overview of the modeling process of solar energy, discussing, among others, several ML approaches, and where short term PV energy forecasting is studied in [15] for both single—and aggregated—PV plants in Germany, with SVR being one of the methods considered.

In this work we will also use SVR [13] for day-ahead prediction and Ridge Regression (RR) [6] for same-day hourly forecast updates. A first reason for the SVR choice is that it is one of the most successful and widely used approaches for non linear modeling. An obvious alternative are Multilayer Perceptrons, but the large dimension of the studied problems implies that, if used, some dimensionality reduction has to be applied to input patterns so that network complexity is manageable. Another reason is the availability of the top quality software LIB-SVM package [2]. Linear models are probably too simple for the fist task but, on the other hand, they are well suited to same-day predictions, given the simplicity of the delay vectors used then as model inputs.

The main goal here is obviously to obtain accurate hourly PV energy predictions. However, this cannot be done directly from NWP forecasts, as they are given only every 3 h. Because of this, our first goal in day-ahead prediction will be to forecast three-hour aggregated energy at UTC hours $h = 3k$, $k = 0, \ldots, 7$; of course, there is no aggregated PV energy at hours 0 and 3, but in Spain there is aggregated energy at hours 6 and 21 from mid spring to mid fall. Given the 1,128 NWP grid used, pattern dimension is thus $2,256 = 2 \times 1,128$. A second goal will be to predict the aggregated daily PV energy over Spain. Here we use the six NWP predictions at UTC hours 6–21 as the input patterns, with now a rather large dimension of $13,536 = 6 \times 2,256$. In any case, we use these 3-hour or daily forecasts as first step towards our main goal of obtaining hourly PV forecasts, which we do by disaggregating the 3-hour and daily predictions. A first idea would be to interpolate them using some physical clear sky radiation model.

---

[1] http://wire1002.ch/fileadmin/user_upload/Documents/ES1002_Benchmark_announcement_v6.pdf

[2] https://www.kaggle.com/c/ams-2014-solar-energy-prediction-contest

A review of some of these models from the point of view of renewable energy is in [10]. However, clear sky models have a very local nature while here we have to disaggregate energy produced all over peninsular Spain and there is no simple way to define a physical model for such a large area. We follow another approach, working with an "empirical clear sky" energy model that we build taking as a basis the hourly maximum normalized PV energy values observed over a number of years. Our results give Mean Absolute Errors (MAE) for the day light hours of about 2–3 % of installed power, with MAE peaks of about 5 % at mid day. While root mean square errors are also widely used, we prefer to work with MAE values as they can be directly translated to energy deviations and, hence, are widely used in the industry. Of course, our errors must be compared with error values obtained elsewhere in the literature but it seems that more research effort has been devoted to short-term PV energy forecasting than to the day-ahead problem and, thus, day-ahead error reference values seem hard to come by.

For same-day PV energy prediction updates, the hourly readings of the PV energy time series is basically the only information available. Because of this and denoting by $\mathcal{E}_{dk}$ the PV energy reading at hour k of a day d, we will simply use a RR linear model where for a given day d, a delay vector $(\mathcal{E}_{d6}, \dots, \mathcal{E}_{dh-1}, \mathcal{E}_{dh})$ is built at hour h to predict PV energy values $\hat{\mathcal{E}}_{dk}$ at hours $k = h + 1, \dots, 20$. The choice of the 6–20 UTC hour interval corresponds to day-light hours in Spain for most of the year.

The paper is structured as follows. We very briefly review SVR and RR in Sect. 2. In Sect. 3 we will describe our 3-hour and daily aggregated PV models and numerically compare their behavior. The empirical clear sky PV energy model is described in Sect. 4 and applied to disaggregate to the hourly level the predictions given by the best 3-hour and daily models; we will also give error comparisons. In Sect. 5 we discuss the RR-based same-day models and the paper ends with a brief discussion and conclusions section.

2   Ridge and Support Vector Regression

Assuming a $N \times d$ dimensional data matrix $\mathcal{X}$ associated to N patterns X_t with dimension d, in linear regression we want a weight vector W such that $X_t \cdot W \simeq y_t$, $t = 1, \dots, N$, or, in matrix notation, $\mathcal{X}W \simeq Y$, where we recall that the rows of $\mathcal{X}$ contain the transposes of the X_t patterns and Y is the N-dimensional output vector with y_t as components. For simplicity we assume the X_t, y_t to have zero mean. In RR the optimal weight W^* is found by minimizing the error function

$$\frac{1}{2}\frac{1}{N}\|\mathcal{X}W - Y\|_2^2 + \frac{\lambda}{2}\|W\|_2^2.$$

In the optimal W^* we have thus a balance between model error and complexity, which we control through the parameter λ. W^* can be found analytically as

$$W^* = (\mathcal{X}^T\mathcal{X} + \lambda I)^{-1}\mathcal{X}^T Y,$$

where I is the $d \times d$ identity matrix. Since $\mathcal{X}^T\mathcal{X}$ is positive semidefinite, $\mathcal{X}^T\mathcal{X} + \lambda I$ is invertible for any $\lambda > 0$. We have to decide on the optimal value of λ, which is usually determined by some form of Cross Validation (CV). While not very powerful to tackle general complex problems, the simplicity of RR makes it suitable for relatively simple problems; we will use it here for same-day, hourly PV energy prediction updates.

The cost function of RR can be written as

$$\sum_t \ell(y_t, W \cdot X_t) + \frac{1}{C}\|W\|^2$$

with $\ell(y, z) = \frac{1}{2}(y - z)^2$ the quadratic loss and $C = \frac{1}{N\lambda}$. Assuming now a non-homogeneous model $W \cdot X + b$, the cost function in SVR [13] is

$$\sum_t [y_t - W \cdot X_t - b]_\epsilon + \frac{1}{C}\|W\|^2 \tag{1}$$

with the loss now being $\ell(y, z) = [y - z]_\epsilon$ and $[v]_\epsilon = \max\{|v| - \epsilon, 0\}$ the ϵ-insensitive loss. We thus allow an ϵ-wide, penalty-free "error tube" around the model. To solve (1), it is rewritten as a constrained minimization problem:

$$\min_{W,b,\xi} \frac{1}{2}\|W\|^2 + C\sum_t(\xi_t + \xi_t^*), \tag{2}$$

subject to the restrictions $W \cdot X_t + b - y_t \geq -\xi_t - \epsilon$, $W \cdot X_t + b - y_t \leq \xi_t^* + \epsilon$ and $\xi_t, \xi_t^* \geq 0$. Its dual problem is then obtained by Lagrangian theory, that yields

$$\min \Theta(\alpha, \beta) = \frac{1}{2}\sum_{t,s}(\alpha_t - \beta_t)(\alpha_s - \beta_s)X_t \cdot X_s +$$
$$\epsilon\sum_t(\alpha_t + \beta_t) - \sum_t y_t(\alpha_t - \beta_t) \tag{3}$$

which has much simpler box constraints $0 \leq \alpha_t, \beta_t \leq C$. The Karush–Kuhn–Tucker (KKT) conditions for problems (2) and (3) can be applied to compute the optimal W^*, b^* of the primal problem from the optimal dual solutions α_t^*, β_t^*, yielding a final model

$$f(X) = f(X, W^*, b^*) = W^* \cdot X + b^* = \sum_t(\alpha_t^* - \beta_t^*)X_t \cdot X + b^*.$$

Notice that $f(X, W^*, b^*)$ is also a simple and perhaps not powerful enough linear model. Since f and (3) only involve dot products, the kernel trick [13] can be used to build f not on the original inputs X but on their extensions $\phi(X)$ to a possibly infinite dimensional Hilbert space. To do so we use an appropriate kernel $K(X, X')$ so that we have $\phi(X) \cdot \phi(X') = K(X, X')$. Thus, we can compute dot products on the extended $\phi(X)$ patterns without actually having to handle them explicitly. Using a Gaussian kernel $e^{-\frac{\|X-X'\|^2}{2\sigma^2}}$ results in a final model

$$f(X) = b^* + \sum_t(\alpha_t^* - \beta_t^*)e^{-\frac{\|X-X_t\|^2}{2\sigma^2}}.$$

Finally, we have to select three parameters, C, ϵ and the Gaussian kernel width σ, which we do again by CV, much costlier now than for RR. In our experiments we will use LIBSVM [2] for SVR and a Matlab's RR implementation.

above. Here we denote the target aggregated production for a day d as E_d, their corresponding predictions as $\hat{E}_d^{D;y}$ for the yearly model and $\hat{E}_d^{D;5m}$ for the 5-month models, and their errors as $e_d^{D;y}$ and $e_d^{D;5m}$ respectively.

An obvious alternative to these direct daily models is to predict the aggregated daily PV energy just as the sum of the 3-hour predictions; we denote now as $\hat{E}_d^{3-D;y}$ the predictions obtained using a yearly model and $\hat{E}_d^{3-D;5m}$ those using the 5-month models. Similarly, we could also consider 3-hour PV energy predictions obtained disaggregating the daily predictions $\hat{E}_d^{D;y}$ and $\hat{E}_d^{D;5m}$ into 3-hour predictions $\hat{E}_{dh}^{D-3;y}$ and $\hat{E}_{dh}^{D-3;5m}$. To do so we need for each day d a sequence $\gamma_h^d, 0 \le h \le 23$, with $\sum \gamma_h^d = 1$ that captures in some way what could be the hourly evolution of PV energy at day d. We will discuss in the next section how to define such a sequence using what we call an empirical clear sky PV energy curve. Assuming available such a table γ_h^d, we can disaggregate a daily energy prediction $\hat{E}^D$ into a 3-hour one at hour h as $\hat{E}_{d,h}^{D-3} = \hat{E}^D \Gamma_h^d$ with $\Gamma_h^d = \gamma_{h-2}^d + \gamma_{h-1}^d + \gamma_h^d$. We disaggregate this way the $\hat{E}_d^{D;y}$ and $\hat{E}_d^{D;5m}$ values into the corresponding 3-hour predictions $\hat{E}_{dh}^{D-3;y}$ and $\hat{E}_{dh}^{D-3;5m}$.

Table 1 summarizes the daily and 3-hour errors of all models considered in this section. As it can be seen, the 5-month models give the best results for both

Table 1. Day-ahead daily, 3-hourly and hourly errors of the yearly and 5-month models over the test months.

Daily model errors

Model	Dec13	Jan14	Feb14	Mar14	Apr14	May14	Ave
$e^{D;y}$	36.03	25.29	31.20	46.81	27.89	23.80	31.84
$e^{3H-D;y}$	44.71	28.36	34.33	48.30	29.80	28.43	35.66
$e^{D;5m}$	30.69	23.41	28.29	**37.12**	**29.22**	**21.84**	**28.43**
$e^{3H-D;5m}$	**22.58**	**23.19**	**27.06**	41.23	31.27	25.15	**28.41**

3-hour model errors

Model	Dec13	Jan14	Feb14	Mar14	Apr14	May14	Ave
$e^{3H;y}$	8.33	**6.17**	6.44	8.71	**6.12**	6.25	7.00
$e^{D-3H;y}$	6.50	6.57	8.18	10.59	7.50	6.51	7,64
$e^{3H;5m}$	**5.28**	6.53	**6.11**	**8.04**	6.68	**5.89**	**6.42**
$e^{D-3H;5m}$	5.77	6.53	7.87	9.52	7.72	6.31	7.29

Hourly model errors

Model	Dec13	Jan14	Feb14	Mar14	Apr14	May14	Ave
$e^{D-H;5m}$	2.28	2.72	3.15	3.79	3.06	2.52	2.92
$e^{3H-H;5m}$	**2.12**	**2.68**	**2.59**	**3.36**	**2.75**	**2.33**	**2.64**

3 Daily and 3-hour Wide Area PV Energy Forecasting

We discuss next the use of SVR to derive daily and 3-hour forecasts of aggregated PV energy. Our data sources are hourly global PV energy values in peninsula Spain from June 1, 2011 to May 31 2014, and NWP forecasts of DSSR and TCC from the ECMWF from December 2012 to May 2014. The NWP period is much smaller, so for testing we will use the six months from December 2013 to May 2014. There is a very small number of missing values in the PV energy record that, nevertheless, is not significant. In particular, there are no missing values in the test period. The input for the 3-hour aggregated energy predictions are the ECMWF forecasts of DSSR and TCC; we recall that pattern dimension is 2, 256. NWP values are given as 3-hour accumulated values for UTC hours 0 3, 6, 9, 12, 15, 18 and 21. Hours 0 and 3 correspond in Spain to night-time all year long and, thus, we will disregard them. Three-hour PV energy value at hours 6 and 21 are also very small or zero from mid fall to mid spring but not so for the rest of the year and we will keep them. Therefore, we will first predict 3-hour accumulated energy for UTC hours 6, 9, 12, 15, 18 and 21 from the corresponding ECMWF forecasts of DSSR and TCC.

We will consider two SVR models. The first one is a single yearly model built using NWP and PV energy values from December 2012 to November 2013 and that we will test in the period from December 2013 to May 2014. We will refer to this as the yearly model. The second approach will be to build month-specific models where the model for month m is built using data from the previous $m - 14$, $m - 13$, $m - 12$, $m - 11$, $m - 10$ months. For instance, the model for May 2014 is built using data from March, April, May, June and July 2013. We will refer to these as the 5-month models.

PV energy varies greatly along the day, possibly because of the different transfer behavior of PV cells. Because of this, we will build the yearly and 5 month models using submodels targeted to specific hours. For the yearly model we will use three different SVR sub-models tailored to UTC hours 6 and 21 (sunrise and sunset), 9 and 18 (mid half day) and 12 and 15 (noon). The 5 month models will be built upon two different SVR sub-models, one of them tailored to hours 6 and 21 and the other to the remaining 9, 12, 15 and 18 h we do this as there are fewer training patterns for them. In our experience this approach yields better results than those obtained using single, all-hour SVR models.

We denote the target aggregated production values for day d and 3-hour h as E_{dh} and their corresponding predictions as $\hat{E}_{dh}^{3;y}$ for the yearly model and $\hat{E}_{dh}^{3;5m}$ for the 5-month model. While there are actually three yearly and two 5-month underlying models, we will report their joint results as those of single models. We denote their errors for day d and 3-hour h as $e_{dh}^{3;y}$ and $e_{dh}^{3;5m}$ respectively.

We will also build similar models to predict now the aggregated daily PV energy generated in Spain. Recall that pattern dimension will then be $13,536 = 6 \times 2,256$, as there are six NWP forecasts for each 3-hour interval from hours 6–21. Again we will consider a yearly model built using the entire year from December 2012 to November 2013 and also the 5-month models described

the daily and 3-hour predictions. Also, the 5-month daily and aggregated 3-hour model essentially tie for daily predictions. However, the 5-month 3-hour models give slightly better results for 3-hour predictions than the disaggregated daily model. Since we normalize productions to installed power, errors are given as percentages of installed power. Thus the average daily best error of about 28.4 % corresponds to a total daily deviation of approximately 1.13 GW and the 6.4 % value of the 3-hour average error to about 256 MW in three hours.

To end this section, we should mention that the SVR parameters C, ϵ and σ have been determined by five-fold monthly stratified CV. By this we mean that we build 5 different random folds for each month with basically six days at each. Instead of retaining just a single C, ϵ, σ set, we will keep the five optimal SVR parameter subsets that we use to build five different SVR models of each one of the types considered, i.e., yearly and 5-month models for daily and 3-hour predictions. Thus, the PV energy forecasts are actually the averages of the ones provided by these models.

4   Hourly PV Energy Forecasts

Recall that once we have the table γ_h^d, $1 \leq d \leq 365$, $0 \leq h \leq 23$, we can disaggregate the daily $\hat{E}_d^{D;y}$ and $\hat{E}_d^{D;5m}$ predictions into the $\hat{\mathcal{E}}_{dh}^{D-H;y}$ and $\hat{\mathcal{E}}_{dh}^{D-H;5m}$ hourly ones simply as

$$\hat{\mathcal{E}}_{dh}^{D-H;y} = \gamma_h^d \hat{E}_d^{D;y}, \ \hat{\mathcal{E}}_{dh}^{D-H;5m} = \gamma_h^d \hat{E}_d^{D;5m}.$$

Similarly, we can disaggregate the 3-hour predictions $\hat{E}_{dh}^{3;y}$ and $\hat{E}_{dh}^{3;5m}$ into the $\hat{\mathcal{E}}_{dh-j}^{3-H;y}, \hat{\mathcal{E}}_{dh-j}^{3-H;5m}$ hourly ones for the hours $h, h-1$ and $h-2$ as

$$\hat{\mathcal{E}}_{dh-j}^{3-H;y} = \frac{\gamma_{h-j}^d}{\sum_{k=0}^{2} \gamma_{h-k}^d} \hat{E}_{dh}^{3H;y}, \ \hat{\mathcal{E}}_{dh-j}^{3-H;5m} = \frac{\gamma_{h-j}^d}{\sum_{k=0}^{2} \gamma_{h-k}^d} \hat{E}_{dh}^{3-H;5m},$$

for $j = 0, 1, 2$. Table 1 also contains the average errors of the 5-month hourly models considered. As it is to be expected, the best hourly results are those given by the best 3-hour model, i.e., the 5-month model. Now the average hourly best error of about 2.64 % corresponds to an hourly error of approximately 105 MW. Table 2 gives the average hourly errors for UTC hours 9–19. The largest error is 4.986 % at hour 13, which corresponds to about 199 MW. Figure 2 compares hourly PV production in the first week of December 2013, February 2014, April 2015 and the last one of May 2014 against the best hourly model and in Fig. 3 are depicted the MAE errors as a percentage of the installed power for the same period of time. As it can be seen, there is a relatively good agreement between production and prediction values and sunrise and sunset are adequately handled.

We discuss next how we obtain the γ_h^d table, which we do using historical PV hourly energy values. The underlying idea is quite simple as we want the interpolating sequence γ_h^d to somehow reflect what would be the evolution of PV energy assuming day d to be a clear sky one. To do so, we first compute for each pair d, h a "maximum energy curve" μ_h^d defined as

$$\mu_h^d = \max_{y,q}\{\mathcal{E}_{d+q,h,y} : -\delta \leq q \leq \delta, \ y\}$$

where δ is some small integer and $\mathcal{E}_{d+q,h,y}$ denotes the energy produced at hour h in day $d+q$ of a year y. In other words, μ_h^d is the maximum of the normalized energy productions registered at hour h and any day in the interval $[d-\delta, \ d+\delta]$ over all years with PV energy production records (in our case from June 2011 to November 2013). We then smooth these μ_h^d values as

$$\rho_h^d = \frac{1}{2n+1}\sum_{-n}^{n}\mu_h^{d+m},$$

re-scale these ρ_h^d values to arrive to

$$\tilde{\gamma}_h^d = \frac{\max_j\{\mu_j^d\}}{\max_j\{\rho_j^d\}}\rho_h^d$$

and finally have

$$\gamma_h^d = \frac{\tilde{\gamma}_h^d / \sum_{h'}}{\tilde{\gamma}_{h'}^d}.$$

We have used the values $\delta = n = 10$ in our experiments. Figure 1 depicts the $\tilde{\gamma}$ curves for May 31, April 1, February 1 and December 1. When compared to actual maximum energy values, it is observed that the $\tilde{\gamma}_h^d$ curves clearly overshoot actual PV production, but since we interpolate using the normalized values γ_h^d, this effect is not important anymore.

5   Same-Day Day Hourly Updates of PV Energy Forecasts

Since NWP forecasts are updated usually only twice a day at around UTC hours 6 and 18, they cannot be used to update previous energy forecasts in an hourly basis. Other sources of information are needed and the most obvious one is the actual hourly production readings up to hour h that, in turn, can be used to obtain new forecast for hours $h+1, h+2, \ldots$ More precisely, for a given day d we may use energy readings $\mathcal{E}_{d6}, \ldots, \mathcal{E}_{dh}$ (we take PV production up to hour 5 as zero) to derive updated predictions $\hat{\mathcal{E}}_{dh+k}$ for the incoming hours $h+k, k = 1, \ldots$

More information can be added, such as the errors $e_{dh-j}, j = 0, 1, \ldots$ of the day ahead predictions, the day ahead NWP forecasts or the day ahead prediction $\hat{\mathcal{E}}_{dh}^{da}$ in use for day d and hour h. However, we will just consider here the simplest approach of building at hour h a linear model $F_k^h(\mathcal{E}_{d6}, \ldots, \mathcal{E}_{dh})$ to approximate $\mathcal{E}_{dh+k}$. In other words, what we want is $\mathcal{E}_{dh+k} \simeq \hat{\mathcal{E}}_{dh+k} = F_k^h(\mathcal{E}_{d6}, \ldots, \mathcal{E}_{dh})$. Of course, the information in $v_h^d = (\mathcal{E}_{d6}, \ldots, \mathcal{E}_{dh})$ will be only relevant for h past enough from sunrise. Thus, the F_k^h models will be only relevant for $h_{SR}^d \leq h \leq h+k \leq h_{SS}^d$, with h_{SR}^d, h_{SS}^d the sunrise and sunset hours of day d. Moreover,

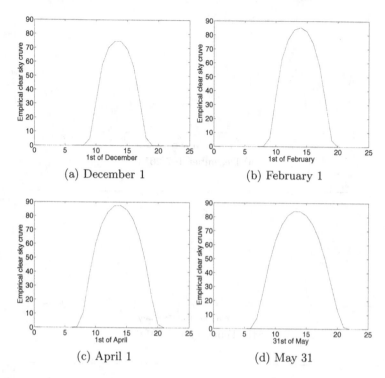

(a) December 1

(b) February 1

(c) April 1

(d) May 31

Fig. 1. Empirical clear sky PV curves for December 1, February 1, April 1 and May 31.

the information in v_h^d will not be of interest for $h \simeq H_{SR}^d$. For these reasons and taking into account sun hours in Spain in the months between December and May, we take UTC hour 6 as sunrise and UTC hour 20 as sunset and thus we shall only consider models F_k^h built starting at hour $h = 9$ and ending their predictions at hour $h + k = 19$.

As mentioned in Sect. 2, we will use RR to build the F_k^h models. Since these models do not need NWP information, we can use the entire period from June 2011 to November 2013 for training and validation purposes. As done before, we have built these models using for training past 5-month periods centered in the month that we use for testing. We estimate the λ_k^h parameter of each RR model using for validation 5-month periods between December 2012 and November 2013. Notice that the maximum number of model parameters is about $18 - 6 + 1 = 13$ while the number of samples is much higher. Thus the λ_k^h have very small values and, in fact, models built using standard linear regression or Lasso give very similar results.

We report our results in Table 2. The top row gives the MAE errors of the best day ahead hourly model (that we recall was given by the 3-hour 5-month SVR models) and the other rows give the average errors over the testing period of

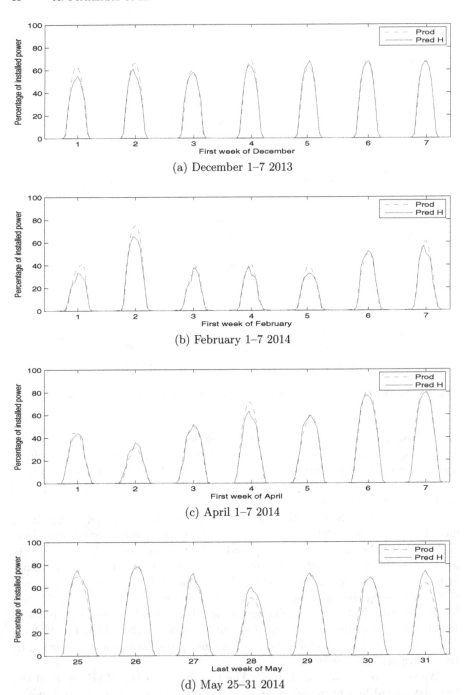

(a) December 1–7 2013

(b) February 1–7 2014

(c) April 1–7 2014

(d) May 25–31 2014

Fig. 2. Hourly prediction (red) vs production (blue) % of installed PV capacity on the weeks starting at December 1, 2013, February 1, April 1, and May 25, 2014 (Color figure online).

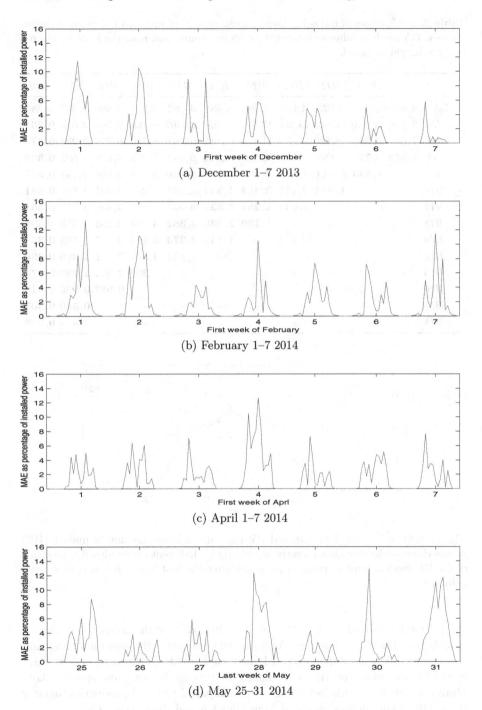

(a) December 1–7 2013

(b) February 1–7 2014

(c) April 1–7 2014

(d) May 25–31 2014

Fig. 3. MAE errors as a % of installed PV capacity on the weeks starting at December 1, 2013, February 1, April 1, and May 25, 2014.

Table 2. MAE errors of intra-day hourly updates as % of installed PV capacity versus the best DA model. Columns represent predicted hours, and rows the hour in which a new prediction is issued.

		H9	*H10*	*H11*	*H12*	*H13*	*H14*	*H15*	*H16*	*H17*	*H18*	*H19*
	DA	4,533	3,552	4,023	4,575	4,986	4,558	3,963	4,204	2,696	1,467	0,384
	H6	9,638	11,570	12,843	13,368	13,573	13,391	12,507	10,326	7,500	3,046	**0,361**
	H7	7,086	9,586	10,820	11,492	11,653	11,509	10,810	8,790	6,358	2,233	**0,372**
	H8	**3,348**	6,772	8,693	9,709	10,127	10,195	9,444	7,485	5,079	1,692	**0,339**
	H9	-	**1,865**	**3,411**	**4,482**	5,338	6,073	6,101	5,785	4,666	1,757	**0,337**
HU	*H10*	-	-	**1,014**	**2,171**	**3,313**	**4,353**	4,992	5,269	4,727	1,691	**0,341**
	H11	-	-	-	**1,044**	**2,254**	**3,322**	4,062	4,655	4,136	1,576	**0,351**
	H12	-	-	-	-	**1,139**	**2,330**	**3,382**	4,305	4,155	1,556	0,403
	H13	-	-	-	-	-	**1,142**	**2,274**	**3,584**	3,707	1,525	**0,334**
	H14	-	-	-	-	-	-	**1,151**	**2,637**	3,361	**1,393**	**0,336**
	H15	-	-	-	-	-	-	-	**1,303**	2,464	1,290	**0,373**
	H16	-	-	-	-	-	-	-	-	0,997	0,946	**0,319**
	H17	-	-	-	-	-	-	-	-	-	0,540	0,307
	H18	-	-	-	-	-	-	-	-	-	-	0,226

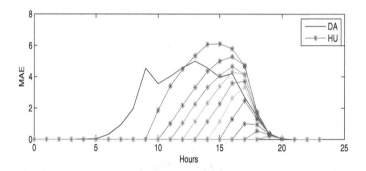

Fig. 4. MAE errors as % of installed PV capacity of same-day hourly update (HU) versus those of the day-ahead hourly models (DA). Different colors identify the errors of the HU models built at consecutive hours after the first one at $h = 9$ (Color figure online).

the F_k^h models with $10 \leq h \leq 19$. We mark in bold face the hours where the F_k^h models beat the day-ahead one. As it can be seen this is the case on a band that contains about 3–4 subdiagonals. Figure 4 captures this behavior, and in Fig. 5 it can be also observed the performance of this model for some specific days. More concretely, in this last image it is compared the real production against the result of the models obtained from hour 9 h and from hour 12 h.

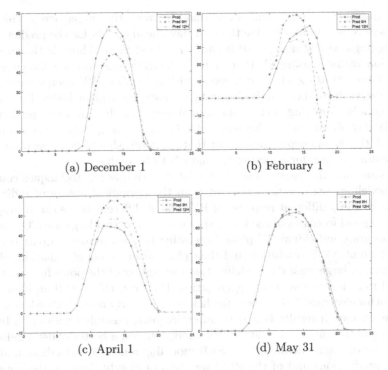

(a) December 1   (b) February 1

(c) April 1   (d) May 31

Fig. 5. Behavior of HU models obtained from hour 9 and from hour 12 for December 1, February 1, April 1 and May 31.

6   Discussion and Conclusions

Wide area prediction of PV energy is an important issue for the Transmission System Operators (TSO) of countries such as Spain where there is a very large number of small and geographically dispersed PV installations. Small individual outputs and dispersion mean that the impact of a single installation is not all that relevant. However, the aggregated effect of all the PV installations is quite important (about 4 GW of peak power in Spain and much higher in other European countries) and an accurate prediction of the global PV output is of a very high interest.

In this work we have addressed day-ahead and intra-day PV energy predictions for peninsular Spain. The inputs in the first, day-ahead case are NWP forecasts from the ECMWF and we derive at day d PV predictions for the entire day $d+1$ at basically two extremes: daily aggregated predictions and individual hourly values. In the second, same-day case, the inputs are the production at a given day up to UTC hour h, $9 \leq h \leq 18$, from which we derive forecasts for UTC hours $h+1, \ldots, 19$. From a ML point of view, both problems are fairly simple as we only have to map input data into accurate predictions. SVR was used for day-ahead models, either as a single model built over an entire year or as month-tailored models built over 5 months, with the target one as the center. These 5-month models give the best day ahead results for the prediction of daily aggregated PV energy and its 3-hour values. Since 3-hour is the smallest resolution available using NWP, to arrive at hourly predictions we disaggregated 3-hour forecasts using what we called empirical clear sky PV energy curves.

To derive intra-day hourly forecasts we have used plain Ridge Regression (RR) models. Applying them from hour 9 onwards, these models' predictions beat the day ahead ones in the next 3–4 h. Given the simplicity of the approach followed, this is an encouraging result that, however, should be improvable possibly adding more information to the models to be built.

In summary, the paper shows how relatively simple ML techniques coupled with modeling considerations derived from the periodical nature of radiation, and the possibly different response of PV cells at different hours can be applied to obtain good forecasts of the PV energy produced over a large area. Of course, while accurate individual PV plant forecasting is much harder (in part because of the current NWP resolution and their physical modeling of radiation-related phenomena), large scale PV modeling takes advantage of the powerful smoothing derived from large area energy aggregation. However, this is still an important problem for electrical TSOs where further work is clearly needed. Besides extending the six month results here over an entire year, possible issues are a better modeling of the empirical clear sky PV energy curves, a better understanding of the nature of the modeling errors (SVR modeling tends to undershoot mid-day energy production) and of the effect on them of cloudy days, or the improvement of the intra-day hourly forecasts. On the other hand, other ML approaches could be used. For instance, in day-ahead predictions we could go from the simplest case of regularized linear models to the more complex one of using, say, Multilayer Perceptrons, possibly after a dimensionality reduction process that provides more manageable input patterns. Moreover, the estimation of forecast uncertainty is also important for TSOs. Theoretical uncertainty estimates are well known for linear models and also do exist for SVR [3,8]. We are working on these and other related issues.

Acknowledgments. With partial support from Spain's grants TIN2010-21575-C02-01 and TIN2013-42351-P, and the UAM–ADIC Chair for Machine Learning. We thank Red Elctrica de Espaa for useful discussions and making available PV energy data.

References

1. Benghanem, M., Mellit, A., Alamri, S.: Ann-based modelling and estimation of daily global solar radiation data: a case study. Energy Convers. Manage. **50**(7), 1644–1655 (2009)
2. Chang, C., Lin, C.: LIBSVM a library for support vector machines. ACM Trans. Intell. Syst. Technol. **2**, 27:1–27:27 (2011). http://www.csie.ntu.edu.tw/~cjlin/libsvm
3. Chu, W., Keerthi, S.S., Ong, C.J.: Bayesian support vector regression using a unified loss function. IEEE Trans. Neural Netw. **15**(1), 29–44 (2004)

4. ECMWF: European Center for Medium-range Weather Forecasts (2005). http://www.ecmwf.int/
5. GFS: NOAA Global Forecast System (2014). http://www.emc.ncep.noaa.gov/index.php?branch=GFS
6. Hastie, T., Tibshirani, R., Friedman, J.: The Elements of Statistical Learning. Springer, Heidelberg (2001)
7. Kleissl, J.: Solar Energy Forecasting and Resource Assessment. Academic Press, New York (2013)
8. Lin, C.J., Weng, R.C.: Simple probabilistic predictions for support vector regression. Technical report, Department of Computer Science, National Taiwan University (2003)
9. Mellit, A., Kalogirou, S.A.: Artificial intelligence techniques for photovoltaic applications: a review. Prog. Energy Combust. Sci. **34**(5), 574–632 (2008)
10. Myers, D.R.: Solar radiation modeling and measurements for renewable energy applications: data and model quality. Energy **30**(9), 1517–1531 (2005)
11. Pedro, H., Coimbra, C.: Assessment of forecasting techniques for solar power output with no exogenous inputs. Sol. Energy **86**, 2017–2028 (2012)
12. Pedro, H., Coimbra, C.: Stochastic learning methods. In: Kleissl, J. (ed.) Solar Energy Forecasting and Resource Assessment, pp. 383–407. Academic Press, New York (2013)
13. Scholkopf, B., Smola, A.: Learning with Kernels: Support Vector Machines, Regularization, Optimization, and Beyond. MIT Press, Cambridge (2001)
14. Ulbricht, R., Fischer, U., Lehner, W., Donker, H.: First steps towards a systematical optimized strategy for solar energy supply forecasting. In: Proceedings of the DARE 2013, Data Analytics for Renewable Energy Integration Workshop, pp. 14–25 (2013)
15. Wolff, B., Lorenz, E., Kramer, O.: Statistical learning for short-term photovoltaic power predictions. In: Proceedings of the DARE 2013, Data Analytics for Renewable Energy Integration Workshop, pp. 2–13 (2013)

The Research on Vulnerability Analysis in OpenADR for Smart Grid

Mijeong Park, Miyoung Kang, and Jin-Young Choi[(✉)]

Department of Embedded Software, Korea University, Seoul, Korea
{mjpark, mykang, choi}@formal.korea.ac.kr

Abstract. Smart Grid has become more important for the efficient use of electric power and the demand reduction. As demand for electric power is increasing continuously despite its limited capacity. The demand reduction in Smart Grid can be achieved through DR (Demand Response) which reduces demand for electric power. In this paper, we analyzed the weaknesses of open source of Open ADR, protocol for Smart Grid DR, using CERT Java secure coding rules. We extracted the violations of rules such as OBJ01-J that means the scope of declaring member variables which should be obeyed in Object-Oriented Programming and IDS00-J that means the validation for input data which should be obeyed in Web environment. By eliminating the weaknesses we could enhance the security of Smart Grid communications.

Keywords: Smart grid · Demand response · OpenADR · Open source · Vulnerability · Security weakness · Secure coding

1 Introduction

Smart Grid is gaining popularity these days since it can solve the problems of increasing demand of electric power, improving the green energy industry at the same time. In addition, the importance of Smart Grid is increasing because of the possibility of blackout caused by the absence of efficient power system management, which can result in economic and national loss.

Smart Grid requires the secure communication environment for the data integrity due to its network-based characteristics. In particular, hacking attacks such as information leakage and manipulation of data can occur, while sending or receiving information between suppliers and consumers. Additionally, Smart Grid is made of combined structure of several systems, with several external connections. As we could see the Stuxnet [1] in 2010 where Iran nuclear power plant got attacked, Smart Grid, a national infrastructure, can be attacked by malicious hackers. It makes the needs of secure system in Smart Grid essential.

In this paper, we suggest several techniques not only to improve security and stability but also to remove systematic weaknesses in Smart Grid by applying secure coding. Also, we propose new secure coding rules for Smart Grid Software, especially when HeartBleed [2] vulnerability of recent OpenSSL Open source has caused a huge amount of loss around the world. It clearly shows that it is necessary to analyze

© Springer International Publishing Switzerland 2014
W.L. Woon et al. (Eds.): DARE 2014, LNAI 8817, pp. 54–60, 2014.
DOI: 10.1007/978-3-319-13290-7_4

vulnerability of Open source used in Smart Grid. We use CERT Java secure coding rules [3] and LDRA Testbed [4] for static analysis based on CERT Java.

The rest of this paper proceeds as follows: Sect. 2 explains the related works on Smart Grid, Secure Coding and analysis tools. Section 3 shows the weaknesses derived from secure coding rules in AMI (Advanced Metering Infrastructure) Software (open source of OpenADR [5]). Section 4 explains conclusion and future works..

2   Related Works

2.1   Secure Coding

CWE (Common Weakness Enumeration) [6] is the standard for measuring software security weaknesses around the world in order to improve security and quality of software. Although it is impossible to eliminate all security weaknesses, we need to put in an effort to minimize those security weaknesses. Secure Coding is the coding standard of source code for developing secure software free of those security weaknesses.

There are a variety of rules in secure coding depending on the characteristics of organizations or fields of study such as Secure Coding Guide developed by the Ministry Of Security and Public Administration in Korea, CERT C/Java developed by the Software Engineering Institute in Carnegie Mellon University and so on.

In this paper, CERT Java is used to analyze open source of OpenADR [8] that is a protocol of Smart Grid, implemented in java.

2.2   Analysis Tool

Quality of software can be improved through static and dynamic analysis of the entire software development process. In this paper, we utilize LDRA TestBed tool [7] for the static analysis that tests source code without executing the program. LDRA's proprietary parsing engine allows us to quickly incorporate new analysis techniques to meet changing standards requirements. Therefore, LDRA Testbed enforces compliance with coding standards and clearly indicates software flaws that might otherwise pass through the standard build and test process to become latent problems. We analyzed whether security weaknesses are removed according to CERT Java secure coding rules on LDRA TestBed.

3   Analyzing Security Weaknesses Based on CERT Java

3.1   Analysis Target

Providers and receivers can transmit and receive data back and forth through the Internet in Smart Grid. In this environment, the critical data such as electric power information and private information can be transferred. During this process, we need to eliminate security weaknesses of the software in order to prevent malicious attacks from hackers.

The analysis target is open source software of OpenADR facilitating data transfer between provider(VTN) and receiver(VEN) in Smart Grid as shown in Fig. 1. There is open source software based on XML for OpenADR.

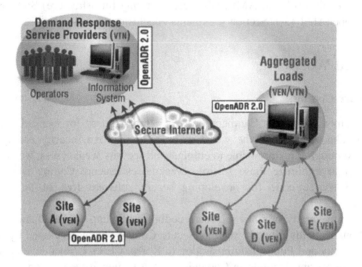

Fig. 1. OpenADR model

We utilize LDRA Testbed 9.4.3 and analyze open source software of OpenADR based on CERT Java secure coding rules. Through this analysis, we detected violations of CERT Java secure coding rules in terms of Static, Complexity and Data Flow.

3.2 Analysis Result

The analysis result on open source software of OpenADR based on CERT Java is shown in Table 1. 'CERT code' means the rule code of CERT Java secure coding and 'Number of Violations' means the number of violations of those rules. From this result, we can see the violations of secure coding rules.

Table 1. The result about violations of secure coding in open source of OpenADR

CERT code	Number of Violations (VEN)	Number of Violations (VTN)	Total	CERT code	Number of Violations (VEN)	Number of Violations (VTN)	Total
IDS05-J	0	0	0	OBJ01-J	627	93	720
IDS06-J	0	0	0	OBJ10-J	404	28	432
EXP00-J	0	2	2	MET02-J	0	0	0
EXP05-J	130	5	135	ERR03-J	2	3	5
EXP06-J	0	0	0	THI00-J	0	0	0
NUM00-J	0	0	0	THI05-J	0	0	0
NUM01-J	0	0	0	FIO02-J	0	0	0
NUM02-J	0	0	0	FIO04-J	0	0	0
NUM07-J	0	0	0	FIO09-J	0	0	0
NUM09-J	0	0	0	MSC01-J	7	7	14
NUM12-J	0	5	5	MSC02-J	0	0	0
NUM13-J	0	0	0				

'OBJ01-J' [9] rule is to declare data members as private and provide accessible wrapper methods. The data members (variables) should be declared as private to prevent the change of data members in unexpected ways. Open source software of OpenADR violated more than 720 rules. Especially, changing data members in unexpected ways can cause serious problems in OpenADR transferring information about electric power usage between providers and receivers.

'OBJ01-J' rule is also related to 'CWE766' [10], the security weakness declaring critical variable as public. 'CVE-2010-3860' [11] is an actual example where the critical data such as user name and directory path were hacked by a remote attacker since they were declared in public.

Violation Example of Fig. 2 is the violated part of 'OBJ01-J' rule in open source of OpenADR. To resolve this security weakness, we need to declare the critical variables as private and provide the public method as shown Modification Example in Fig. 2. By applying this method we can prevent direct access to these variables.

Violation Example

```
public class ContentType implements Serializable, Equals, HashCode, ToString
{

    protected List<Object> content;
    protected String src;

    private final static long serialVersionUID = 1L;
```

Modification Example

```
public class ContentType implements Serializable, Equals, HashCode, ToString
{

    private List<Object> content;
    private String src;

    public void setSrc(String value) {
        // add codes for validating input value
        this.src = value;
    }
}
```

Fig. 2. The solution of OBJ01-J rule violation in open source of OpenADR

'EXP00-J' [12] rule is not to ignore the values returned by methods. The method returns the value to notify success/failure or to update the value. If we ignore or don't handle the returned value from the method properly, it can cause security problems. Therefore, we should handle the return value of the method in a proper manner to prevent the security problems in advance.

'EXP00-J' rule is also related to 'CWE252' [13], the security weakness caused by not checking the returned value from the method. 'CVE-2010-0211' [14] is an actual case of DoS (Denial of Service) attack resulting from the violation of 'EXP00-J' rule in OpenLDAR (open source of LDAP).

Violation Example in Fig. 3 is the violated part of 'EXP00-J' rule in open source software of OpenADR. To resolve this security weakness, we should handle the return value from the method properly as shown in Modification Example of Fig. 3.

Violation Example

```
protected Map<String, Collection<CacheOperationContext>> createOperationContext(
        Collection<CacheOperation> cacheOperations, Method method,
        Class<?> targetClass, HttpServletRequest request) {

    Map<String, Collection<CacheOperationContext>> map = new LinkedHashMap<String, Collection<CacheOperationContext>>(3);

    Collection<CacheOperationContext> cacheables = new ArrayList<CacheOperationContext>();
    Collection<CacheOperationContext> evicts = new ArrayList<CacheOperationContext>();
    Collection<CacheOperationContext> updates = new ArrayList<CacheOperationContext>();

    Object[] args = findArgs(request, method);
```

Modification Example

```
protected Map<String, Collection<CacheOperationContext>> createOperationContext(
        Collection<CacheOperation> cacheOperations, Method method,
        Class<?> targetClass, HttpServletRequest request) {

    Map<String, Collection<CacheOperationContext>> map = new LinkedHashMap<String, Collection<CacheOperationContext>>(3);

    Collection<CacheOperationContext> cacheables = new ArrayList<CacheOperationContext>();
    Collection<CacheOperationContext> evicts = new ArrayList<CacheOperationContext>();
    Collection<CacheOperationContext> updates = new ArrayList<CacheOperationContext>();

    Object[] args = findArgs(request, method);

    if(args != null){
        // add codes after checking return value
    }
```

Fig. 3. The solution of EXP00-J rule violation in open source of OpenADR

After modifying open source of OpenADR according to CERT Java Secure Coding Rules, we confirmed that the program of open source of OpenADR was executed successfully.

3.3   Analysis Result Through Setting up the Test Environment

Smart Grid requires the web environment since it transfers data through the common networks between providers and receivers. For this reason, complying with 'Input Validation and Data Sanitization (IDS)' rule of CERT Java secure coding is essential. The set of IDS rules includes a total of 14 kinds of rules, 'IDS00-J (Sanitize untrusted data passed across a trust boundary)' rule and IDS11-J (Eliminate non-character code points before validation) rule are especially important among them.

We set up the test environment to find the security weaknesses that cannot be detected by the LDRA tool. Figure 4 is the result of testing Request and Response using open source software of OpenADR between providers and receivers. This test is the example of the wrong input value intentionally made. In other words, it enters XML tag as a value of requestID field, resulting in unexpected value as shown in Fig. 4 Response. This result can cause huge damage where we are not able to control correct usage of electric power under emergency in Smart Grid.

We can prevent the wrong intentionally made input value as shown in Fig. 4 through the specific pattern that can be used as input values as Fig. 5. Therefore it needs to establish the specific pattern for Smart Grid and to perform the validation of input values.

We verified the importance of complying with 'IDS00-J (Sanitize untrusted data passed across a trust boundary)' rule when utilizing open source software of OpenADR based on the web environment by setting up the test environment. In this respect,

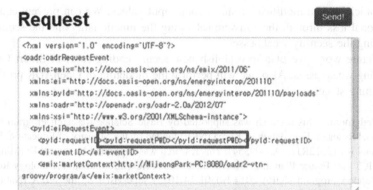

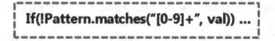

Fig. 4. The example of the wrong input value

```
If(!Pattern.matches("[0-9]+", val)) ...
```

Fig. 5. The example of the specific pattern

verification of input values is needed, which can be implemented by restriction using the specific pattern.

4 Conclusion

This paper analyzed security weaknesses in open source software of OpenADR, which is DR protocol for demand reduction of Smart Grid. We derived several secure weaknesses of OpenADR, which violate secure coding rules according to CERT Java, by using the LDRA Testbed tool. In addition, we conducted experiments on the web environment in order to detect secure weaknesses, which could be found when we utilize the LDRA Testbed tool. As a result, we drive several secure weaknesses that allow malicious data modification and wrong input values. We can prevent economic and national loss through the software satisfying the functionality and the security by eliminating the security weaknesses.

As future work, we plan to establish new secure coding rules of Smart Grid for eliminating weaknesses. Also, we will perform the modeling of OpenADR protocol at its designing stage.

Acknowledgment. This research was supported by Basic Science Research Program through the National Research Foundation of Korea (NRF) funded by the Ministry of Education, Science and Technology (2012R1A1A2009354). This research was supported by the MSIP (Ministry of Science, ICT and Future Planning), Korea, under the ITRC (Information Technology Research Center) support program (NIPA-2014-H0301-14-1023) supervised by the NIPA (National IT Industry Promotion Agency).

References

1. Kushner, D.: The real story of stuxnet. IEEE Spectr. **50**(3), 48–53 (2013)
2. CVE, CVE-2014-0160. CVE.MITRE (2014)
3. Joe McManus, M.G.R.: The CERT Oracle Secure Coding Standard for Java. CERT (2014)
4. Hennell, M.: LDRA Testbed and TBvision. LDRA (1975)
5. OpenADR Alliance, OpenADR 2.0 Profile Specification A Profile. OpenADR Alliance (2011)
6. CWE, Common Weakness Enumeration. CWE.MITRE (1999)
7. LDRA, LDRA Getting Started Tutorial. LDRA Software Technology
8. McParland, C.: OpenADR open source toolkit: developing open source software for the smart grid. In: IEEE Power & Energy Society General Meeting (2011)
9. CERT, OBJ01-J. Declare data members as private and provide accessible wrapper methods. CERT (2012)
10. CWE, CWE-766: Critical Variable Declared Public. CWE.MITRE (2014)
11. CVE, CVE-2010-3860. CVE.MITRE (2010)
12. CERT, EXP00-J. Do not ignore values returned by methods. CERT (2014)
13. CWE, CWE-252: Unchecked Return Value. CWE.MITRE (2014)
14. CVE, CVE-2010-0211. CVE.MITRE (2010)

Improving an Accuracy of ANN-Based Mesoscale-Microscale Coupling Model by Data Categorization: With Application to Wind Forecast for Offshore and Complex Terrain Onshore Wind Farms

Alla Sapronova[1(✉)], Catherine Meissner[2], and Matteo Mana[2]

[1] UniResearch Ltd., University of Bergen, Bergen, Norway
alla.sapronova@uni.no
[2] WinSim Ltd., Tonsberg, Norway
{catherine,matteo}@windsim.no

Abstract. The ANN-based mesoscale-microscale coupling model forecasts wind speed and wind direction with high accuracy for wind parks located in complex terrain onshore, yet some weather regimes remains unresolved and forecast of such events failing. The model's generalization improved significantly when categorization information added as an input. The improved model is able to resolve extreme events and converged faster with significantly smaller number of hidden neurons. The new model performed equally good on test data sets from both onshore and offshore wind park sites.

Keywords: Wind speed forecast · Data categorization · Artificial neural network

1 Introduction

It is important to forecast an accurate wind energy yield as the wind energy production in many countries became a large part of the grid supply. The variability of wind energy remains main challenge for grid engineers, wind park owners, and electricity market players: i.e. practitioners, operating in the field where accurate forecast is required (often obliged by governmental regulations) in the range from minutes to 72 h ahead.

To forecast a wind speed and directions, the most important components for energy yield prediction, Numerical Weather Prediction (NWP) mesoscale models at a coarse resolution of tens of kilometers are commonly used. For site-specific accurate forecast a coupling between mesoscale model output and observation at wind park location is crucial. This coupling requires modeling of the wind flow near the ground, where terrain roughness and complexity affect the flow at microscale. That involves predictive modeling of nonlinear multi-variable function in the environment where explicit physics-based models either have limited application or not available.

Recently a new mesoscale-microscale coupling model was proposed [1]. The model is based on artificial intelligence methods (i.e. unsupervised machine learning)

© Springer International Publishing Switzerland 2014
W.L. Woon et al. (Eds.): DARE 2014, LNAI 8817, pp. 61–66, 2014.
DOI: 10.1007/978-3-319-13290-7_5

and uses NWP forecasts on temperature, pressure, relative humidity, wind speed, and wind direction) to issue a site-specific forecast. Trained against historical observations of wind speed/wind direction, the model forecasts mean hourly wind speed/wind direction to be further utilized as an input for flow model (CFD) at finer scales where roughness and wake effects are taking into account [2].

On the test data sets the model predicted the wind speed in a very satisfactory manner with MSE = 1.8 (m/s) for one-hour ahead prediction. Even though such accuracy found to be superior to that based on polynomial fittings as well as ARMA models, the detailed statistics on model performance shows that for some weather regimes MSE was significantly higher than others.

In this work, the model proposed in [1], is developed further to achieve the same level of forecast accuracy for the entire data set, including weather regimes that original model was not able to catch. Improved model uses data categorization approach, when information obtained from categorization of a single variable (wind speed) was supplied to ANN input in addition to above mentioned NWP variables. Improved model showed lower MSE = 1.1 (m/s) for wind speed prediction on the test data sets, resolving all weather regimes at the same level of accuracy.

2 Model Selection

Machine learning (ML) based models' generalization arising from model's ability to find similarity in training data that usually consists of continuous numeric data. Since numbers are rarely exactly the same from one example to the next, the model can fails in selecting the margins for identical properties. In this case, the generalization can be improved by classification. For example, a combination of ML methods like self organizing maps (SOM) and feed-forward neural networks (FF ANN) can lead to forecast improvement [3]. Unfortunately, if the training data either limited or incomplete bringing SOM for forecast improvement became a challenge itself.

Also, some examples in the training set, that normally treated equally can vary in reliability or carry less critical information about the target than the others, or can even carry wrong information. E.g. wind speed measurement device on site can fail for variety of reasons, and sometimes with greater error in specific wind speed range(s).

According to the intrinsic margin, physical nature of the problem, etc., training data-sets can be grouped into several discrete categories. The discrete categories will allow identical category values to be treated in the same manner.

One logical approach is to categorize numeric data, a wind speed in this case, similar to typical human concepts (e.g. "calm wind", "fresh breeze", "strong wind", etc.) and then try to generalize. The task can be defined as trying to divide numeric fields into appropriate sets of categories.

While most of the researches done on learning from examples has not been concerned with categorizing numeric data, some experimental results [4] show that choice of methods for categorization is irrelevant to the generalization improvement. Methods, quite different by nature, give similar and reasonable results with real-world data-sets and all lead to improved generalization.

3 Data

In this work the data from ERG Wind Farm located in Molise in central Italy was used. The farm location is in a wide area between the town of Ururi and Montelongo, situated in complex terrain of different ridges and is about 10 km wide. The wind farm layout is composed by 20 Vestas V90 turbines.

The anemometer used as a reference, located in a central position of the wind farm at 30 m height.

The measurement data sets (anemometer registered values) start 1/04/2010 and end 1/03/2012. The time series almost two years long thus allow the entire year of data to be used in the training set and the rest period is used to validate the performance of the trained network.

The provided by Meteogroup, NWP data is a forecast of five days delivered four times a day. The time series start 1/04/2010 and end 6/05/2012 to keep a concurrent period with measured data. The data forecasted at two different heights: 80 m and 30 m; at both heights the variables forecasted are wind speed, wind direction, temperature, pressure, density and relative humidity.

Prior usage, the anemometer registered values were cleaned from invalid data and all the events with wind speed under 0.5 m/s were excluded. Then the measured and forecasted values were pre-processed and normalized. Finally, the data was fed into feed-forward neural network as described in Fig. 1: the wind data of the anemometer was used as target for FF ANN training against NWP.

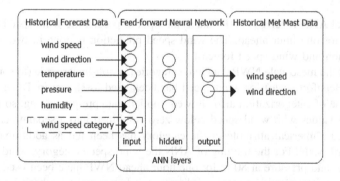

Fig. 1. Data usage for FF ANN training and validation

This model output further utilized for a correction that connects the raw forecast to the measured wind data as showed in Fig. 2. This correction is used for the energy yield calculation in WindSim software.

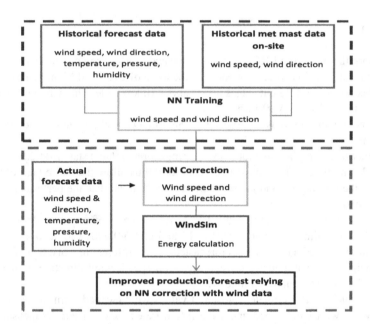

Fig. 2. The schema illustrates the employment of trained FF ANN for improved energy yield forecast

4   Results

The artificial ANN receives NWP data as an input to predict the wind speed inside the wind farm for one hour ahead. The wind speed prediction is done in two steps: data categorization and wind speed forecast.

To split the mesoscale NWP data into categories, a non-linear scale (to some extend similar to Beaufort scale) for wind speed has been used as shown in Table 1.

This type of categorization also allowed implicit data pre-processing, so that failed wind speed values with wind speed below zero combined into a separate category.

Obtained numerical attribute of the category has been fed as additional input to feed-forward ANN. For the training of ANN the wind speed category, wind direction, temperature, and pressure at 80 m for one hour ahead NWP have been used as inputs.

On-site registered wind speed and direction at forecasted time have been used as desired output for training or for model test and validation.

Table 1. Wind speed ranges and corresponding numerical attributes used for categorization.

Wind speed range, m/s	<0.5	0.5–1	1–3	3–7	7–10	10–15	15–20	20–25	>25
Category	failed record	wind calm	light breeze	gentle breeze	fresh breeze	strong breeze	near gale	gale	cut off speed
Numerical attribute	1	2	3	4	5	6	7	8	9

The performance of proposed model containing categorization values has been compared to (a) non-categorization model and (b) classification model, where input variables (excluding category variable) have been enhanced by classification identification obtained from various SOM. (In latest case, NWP parameters, like temperature, wind speed, wind directions, have been quantized by SOM and a separate ANN was used to find the correlation between inputs and desired outputs). Best SOM classification has been achieved by 2×3 matrix.

For non-categorization approach, two models were created (referred as model I and model II). In model I two inputs (wind speed and wind direction) from latest NWP run were used for ANN input. In model II 54 inputs, containing 6 variables (temperature, humidity, stability, pressure, wind speed, wind direction) issued by latest 6 NWP runs (1–72 h ago) were used for ANN input. The summary for performance of all the above mentioned models shown in Table 2.

The proposed categorization approach has been tested on data from offshore wind farm: the new FF ANN was trained on the data a wind park consisting of 90 turbines and located in North Sea. The data contains hourly met mast reading and NWPs for years 2005–2007. The categorization model performance was compared with non-categorization model (both models had four inputs: wind speed category/wind speed NWP data, wind direction, temperature, and pressure and were trained against measured data). The comparison shows that categorization model performs equally good for offshore and onshore wind forecast: significant improvement in wind speed forecast for categorization model (RMSPE 3.2 % vs 5.2 %) was observed with number of hidden neurons lowered from 15 in non-categorization model to 7 in categorization model.

Table 2. Comparision of different FF ANN models' performance

Model	With categorization	Non-categorization I	Non-categorization II	Classification
ANN architecture with winning performance	4 input, 9 hidden, 1 output neurons	2 input, 30 hidden, 1 output neurons	36 input, 60 hidden, 1 output neurons	4 input, 30 hidden, 1 output neurons
Training time required to reach 0.01 % training error, number of iterations rounded to thousands	20,000	240,000	2,185,000	438,000
Mean absolute percentage error	2.4 %	4.6 %	4.1 %	5.0 %
Root mean square percentage error (RMSPE)	5.4 %	9.8 %	8.8 %	10.4 %
Coefficient of determination (R^2)	0.83	0.74	0.77	0.69
Correlation	0.87	0.62	0.79	0.67

5   Conclusions

It is shown that ANN, that was previously used successfully for mesoscale-microscale models coupling can be improved significantly by adding categorization information.

It is observed that model with added categorization information has nearly twice lower RMSPE than regular model (5.4 % vs 9.8 %). Surprisingly, adding SOM classification output to the model input slightly lowered the generalization ability of the network (10.4 % vs 9.8 %). This can be explained in the way that after SOM quantized the entire data-set into 2 × 3 classes less examples became available for ANN training associated with each class. Assuming that enough data samples can be obtained to successfully train SOM-produced ANNs, it is important to note, that categorization model requires almost three times less hidden neurons (9 vs 30). Therefore data categorization more successfully lowers the data dimensions comparing to classification made by SOM. While, data classification is also valuable and shall not be underestimated. E.g in the current work, the visualization of SOM output provides additional information on data patterns and can lead to more successful choice of categories. Also classification patterns evolution can be used to study the dynamics of wind flow inside the wind farm.

Acknowledgment. This work is sponsored by Norwegian Research Council, project ENER-GIX, 2013–2014.

References

1. Sapronova, A., Meissner, C., Mana, M.: Mesoscale-microscale coupled model based on artificial neural network techniques for wind power forecast. Poster at EWEA Offshore 2014, PO ID 385
2. Mana, M.: Short-term forecasting of wind energy production using CFD simulations. Poster at EWEA 2013 (2013)
3. Toth, E.: Combined use of SOM-classification and Feed-Forward Networks for multinetwork streamflow forecasting. Geophysical Research Abstracts, vol. 11, EGU2009-11962 (2009)
4. Li, L., Pratap, A., Lin, H.-T., Abu-Mostafa, Y.S.: Improving generalization by data categorization. In: Jorge, A.M., Torgo, L., Brazdil, P.B., Camacho, R., Gama, J. (eds.) PKDD 2005. LNCS (LNAI), vol. 3721, pp. 157–168. Springer, Heidelberg (2005)

PowerScope: Early Event Detection and Identification in Electric Power Systems

Yang Weng[1]([✉]), Christos Faloutos[2], and Marija Ilic[1]

[1] Department of Electrical and Computer Engineering, Carnegie Mellon University,
Pittsburgh, PA 15213, USA
yangweng@andrew.cmu.edu, milic@ece.cmu.edu
[2] School of Computer Science, Carnegie Mellon University,
Pittsburgh, PA 15213, USA
christos@cs.cmu.edu

Abstract. This paper is motivated by major needs for fast and accurate on-line data analysis tools in the emerging electric energy systems, due to the recent penetration of distributed green energy, distributed intelligence, and plug-in electric vehicles. Instead of taking the traditional complex physical model based approach, this paper proposes a data-driven method, leading to an effective early event detection approach for the smart grid. Our contributions are: (1) introducing the early event detection problem, (2) providing a novel method for power systems data analysis (PowerScope), i.e. finding hidden power flow features which are mutually independent, (3) proposing a learning approach for early event detection and identification based on PowerScope. Although a machine learning approach is adopted, our approach does account for physical constraints to enhance performance. By using the proposed early event detection method, we are able to obtain an event detector with high accuracy but much smaller detection time when comparing to physical model based approach. Such result shows the potential for sustainable grid services through real-time data analysis and control.

Keywords: Power systems · Smart grid · Early event detection · Data mining · Nonparametric method · Machine learning

1 Introduction

Regarded as a seminal national infrastructure, the electric power grid provides a clean, convenient, and relatively easy way to transmit electricity to both urban and suburban areas. Therefore, it is critical to operate the electric power systems in a reliable and efficient way. However, the grid often exhibits vulnerability to penetrations and disruptive events, such as blackouts. Interruptions of electricity service or extensive blackouts are known contributors to physical hardware

Yang Weng is supported by an ABB Fellowship.

© Springer International Publishing Switzerland 2014
W.L. Woon et al. (Eds.): DARE 2014, LNAI 8817, pp. 67–80, 2014.
DOI: 10.1007/978-3-319-13290-7_6

damages, unexpected interruption of normal work, and subsequent economic loss. To monitor potential problems [1], hidden state information is usually extracted from redundant measurement data in the Supervisory Control And Data Acquisition (SCADA) system using static state estimation (SE) [2]; the results of SE are then used for on-line assessment of abnormal event [3] detection.

However, such a static model-based event detection method often causes a significant gap in performance between the state-of-the-art methods and the desired informative data exploration. In particular, with recent and ongoing massive penetration of renewable energy, traditional methods are unable to track and manage increasing uncertainties inherently associated with these new technologies [4]. Further, President Obama's goal of putting one million electric vehicles on the road by 2015 will also contribute to the grid architecture shift. Because of the unconventional characteristics of these technologies, faster and more accurate event detector must be conducted for the robust operation of smart grid.

On the other hand, more sensors [5,6], high performance computers, and storage devices are deployed over the power grid in recent years, which creates an enormous amount of data that is unbelievable in the past. Utilizing these data can be regarded as a perfect opportunity to test and resolve the problem stated above. Notably, learning from the data to deal with uncertainties has been widely recognized as a key player in achieving the core design of Wide Area Monitoring, Control and Protection (WAMPAC) systems, which centers on efficient and reliable operations [7].

In this paper, we propose a fast event detection method caller early event detector. We will use historical data to help predict the complete event sequence based on a partial measurement sequence. Such short sequence may seem to be negligible from an system operator's field operation, but it contains rich information about what is going on next based on the historical data. As processing historical data requires huge computational time, we start with analyzing the speedup possibility by looking into power grid data structure via Singular Value Decomposition (SVD) [8] and Independent Component Analysis (ICA). Such a method is named PowerScope in this paper. Based on our understanding of the possibility of dimension reduction and independent power flow component, a data-driven early event method is then proposed in a nonparametric model [9]. To embed power flow equation inside our detector, a polynomial kernel is used in the regression process. Our simulation over standard IEEE test cases showed that one can detect events much faster rather than waiting to see the complete event shows up. Such an algorithms can form the basis for smart grids analysis by converting large complex system analysis into small physically meaningful independent components to enhance system operation in a predictable ways. Finally, two more applications based on PowerScope are proposed: (1) geographical-plot for visualization and (2) greenness for grid evaluation.

The rest of this paper is organized as follows: In Sect. 2, we review the power systems model and define the problem. In Sect. 3, we first conduct a data analysis for power systems; subsequently, we propose to use independent components with kernel ridge regression to conduct efficient early event detection. Two other applications based on data analysis are described in Sect. 4. In Sect. 5, we illustrate the simulation results. Finally, we conclude our analysis in Sect. 6.

2 Power Systems Data and Problem Definition

2.1 Power Systems Data

A power grid is defined as a physical graph $G(V, E)$ with vertices V that represent the buses (generators and loads), and edges E that represent transmission lines and transformers. The graph of the physical network can be visualized as the physical layer in Fig. 1a. Over this physical network, there is a cyber layer, which collects physical measurements and communicates them back to SCADA Center for further analysis. As an illustration, an artificial three-bus system resembling Pittsburgh, Philadelphia, and Washington D.C. is illustrated in Fig. 1b. The black color represents the physical connection. The blue and red colors represent the measurement data to be communicated and analyzed.

The measurement model of static AC power system SE is expressed as follows [2]:

$$z_i = h_i(\boldsymbol{v}) + u_i, \tag{1}$$

where the vector $\boldsymbol{v}$

$$\boldsymbol{v} = (|v_1|e^{j\delta_1}, |v_2|e^{j\delta_2}, \cdots, |v_n|e^{j\delta_n})^T \tag{2}$$

represents the power system states, where $|v_1|$ is the voltage magnitude at bus one and δ_1 is the phase angle at bus one. u_i is the ith additive measurement noise, which is assumed to be independent Gaussian random variable with zero mean, i.e., $\boldsymbol{u} \sim \mathcal{N}(\boldsymbol{0}, \Sigma)$, where Σ is a diagonal matrix, with the ith diagonal element σ_i^2. z_i is the ith telemetered measurement, such as power flows and power injections. Thus, $h_i(\cdot)$ is a nonlinear function for the ith measurement.

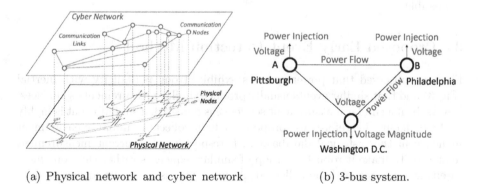

(a) Physical network and cyber network (14-bus system).

(b) 3-bus system.

Fig. 1. Power grid (Color figure online).

2.2 Problem Definition

Instead of conducting model based analysis over a single time slot, we aim at performing historical data-driven early event detection and identification for power systems in this paper. We consider a sequence of historical SCADA measurements obtained every 2 s (or shorter, when faster devices such as Phasor Measurement Units are available.). These measurements are represented as a discrete time sequence $\{z_1, z_2, \cdots, z_k, \cdots, z_K\}$. For instance each column of Fig. 2 represents a measurement set at a time slot. Besides, historical events, such as event with type A, are associated with corresponding time slots. We assume that a new event can only be identified after a relative long period of time. The problem we try to solve here is to use the historical data to detect and identify a new event as soon as possible. The formal problem definition is:

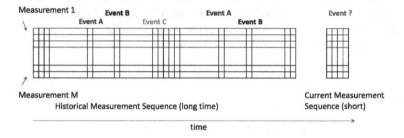

Fig. 2. Problem definition.

- Problem Name: Early Event Detection and Identification
- Given:
    - a sequence of long historical measurements $\{z_1, z_2, \cdots, z_k, \cdots, z_K\}$.
    - a list of events associated with time stamps.
- Find: a data driven event detector and identify the event type as fast as possible.

3 Proposed Early Event Detection Method

First, we observed that power systems exhibit strong periodicity with inertial (Fig. 3) and that similar events usually produce similar measurements sequences. Reversely, if a similar measurement sequence is detected, a similar event is highly likely to incur. Therefore, we propose to first collect a similar measurement sequence in the historical database with respect to the current measurement sequence. To make it robust, a group of similar sequences rather than one measurement sequence shall be collected. A direct way is to report an event, once we find that similar historical measurement sequences are frequently associated with an (abnormal) event. However, such a method not only need to wait until whole event completes, but also requires long exhaustive search time.

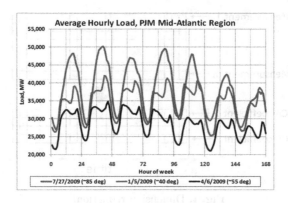

Fig. 3. Power system pattern.

To reduce the nearest neighbor sequences search time, we propose to use the ICA to map the data into lower dimensional space as illustrated in Fig. 4. To conduct the detection process earlier, we employ prediction techniques based on kernel ridge regression. Thereafter we use the current measurement data together with the predicted future measurements for event identification.

Therefore, the proposed algorithm can be described as:

– Step 1: use ICA to map the data onto low dimensional space.
– Step 2: nearest neighbor sequences search over low dimensional space.
– Step 3: kernel ridge regression: predict future measurements with historical nearest neighbor sequences.
– Step 4: use the current measurement sequence in combination with the predicted measurement sequence to find another nearest neighbor sequence in the database.
– Step 5: declare the event associated with the historical neighbor sequence.

In below, we introduce the dimension reduction, nearest neighbor search, and kernel ridge regression for prediction.

3.1 Dimension Reduction-PowerScope

Singular Value Decomposition: The power system has enormous amount of buses. Therefore, directly working on the grid data is prohibitive due to the curse of dimensionality. However, as the electrical power system exhibits periodicity (i.e. Fig. 5a), we expect the measurement data to be highly clustered as to creating the possibility for great dimension reduction for measurement data. In this paper, we propose to conduct SVD over the historical data to see if dimension reduction is possible. We collect one year data of a power grid. As the data is collected every 5 min, we obtain $K = 47520$ data points. The power grid has 300 buses, representing 300 nodes connected via 460 branches. In each data point, we have multiple measurements distributed on the buses (nodes) and transmission

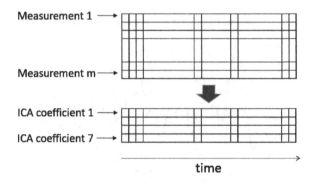

Fig. 4. Dimension reduction.

lines (branches). In our analysis, we have different measurements, namely the power flow measurements and power injection measurements. They contributed 1073 measurements per time slot over a period of one year, which are used to form the historical measurement matrix $Z = [z_1, z_2, \cdots, z_K]$.

The SVD decomposition below is then applied to the historical measurement matrix Z.

$$Z = U \times S \times V', \tag{3}$$

where the diagonal entries of S are known as the singular values of S. U and V are unitary matrices [10]. By plotting the singular values of Z in Fig. 5a with a $\log - \log$ scale, only 8 significant singular values occurred. Besides, in Fig. 5b, we show the result of mapping the historical data onto certain two dimensional features (two left-singular vectors of U) associated with significant singular values. This verifies our expectation of clustered data points.

3.2 Independent Component Analysis

To further analyze the pattern of the power grid data, we conduct Independent Component Analysis (ICA) on the same data source as used in the analysis of Sect. 3.1. Differently, the same data is now represented by a new matrix B. Each row now represents a daily power injection, or power flow measurements within a particular day. In ICA, the task is to transform the observed data B into maximally independent components s using a linear static transformation W as $s = Wb$ [11]. As a result, the components b_i are generated as a sum of the independent components s_k, $k = 1, \cdots, n$:

$$b_i = a_{i,1}s_1 + \cdots + a_{i,k}s_k + \cdots + a_{i,n}s_n, \tag{4}$$

weighted by the mixing weights $a_{i,k}$.

To implement ICA into our analysis, we use the software Fastica Package for MATLAB [12]. As an illustration, Fig. 6a shows a result of applying the Fastica package over two daily signals. As the two signals exhibit high power

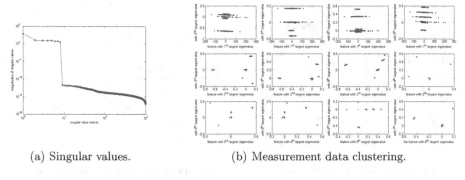

(a) Singular values.   (b) Measurement data clustering.

Fig. 5. Singular value decomposition (SVD).

usage during the day, independent component analysis converted them into signals with in-common and distinctive features on the right of Fig. 6a. The blue
solid line represents basic power consumption, which corresponds to high power
usage/generation in day time. The green dash line represents randomness on top
of the blue solid line. Figure 6b shows the result of applying Fastica to matrix B.
The top two sub-figures represent two peaks of power consumption hours, 8 a.m.
and 2 p.m. The first sub-figure in the second row represents high power usage
in the day time. "-6" on the vertical coordinate represents high power usage in
day time. The second figure in the second row represents two power flow changes
happen at 6 a.m. and 6 p.m. The rest three sub-figures depict randomness within
the grid. We call such a decorrelation process of the power system data "Power-
Scope". In the following, the seven independent components are used data-driven
early event detection, because they are compact representations of the physically
meaningful signals, which can be used for computationally-tractable inferences.
More PowerScope-based applications will be introduced in Sect. 4.

3.3   Nearest Neighbors Search

For simplicity, in this subsection, we use formula (5) to find K nearest neighbor
sequences in the historical database.

$$\hat{s} = \arg \min_{|s|=p} d(s) = \sum_{k \in s} ||z_{current}^{seq} - z_k^{seq}||_2^2, \; k \le K, k \in \mathcal{N} \tag{5}$$

i.e., minimizing sum distance function $d(s)$. Here K is the number of total data
points in the database. $\mathcal{N}$ represents the set of natural numbers. k represents
a particular index for a data point. As a result, z_k^{seq} indicates the measurement sequence starting from the time slot of index k with a length b. Finally, p
indicates the cardinality of the set s. Essentially, during the search step, the algorithm simply looks for an index set s with p elements which represents a group
of measurement sequences that have nearer distance to the current sequence
$z_{current}^{seq}$.

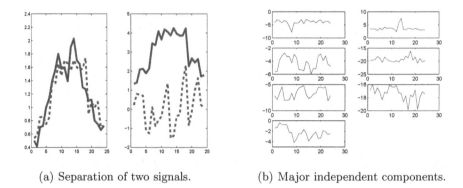

(a) Separation of two signals.   (b) Major independent components.

Fig. 6. Independent component analysis (Color figure online).

3.4 Kernel Ridge Regression

After obtaining the minimum distance index set s, this subsection aims to use the complete historical measurement sequences as well as the current partial measurement sequence to obtain a good prediction for subsequent future sequences. Then we have an understanding what is the potential event that is going on, rather than waiting to the end of it. Afterwards, the complete sequence is used to identify the associating event and report earlier. Such a process can be illustrated in Fig. 7. To simplify the computation process, each historical measurement sequence is vectorized into a single column vector. As a result, each slice of matrix in Fig. 7 will be represented as a vector. For example, the ith historical measurement sequence z_i^{seq} on the top left will be denoted as a vector w_i. The ith historical measurement sequence on the top right will be vectorized as a vector y_i. The current sequence on the bottom left $z_{current}^{seq}$ is vectorized by a vector $w_{current}$. The future measurement sequence to be predicted is vectorized by a vector y_{future}.

Ridge Regression: We explain the process by first considering the Normal model below, which is a popular discriminative model with unknown hyperparameters q and Σ_d:

$$y|w : N(q^T w, \Sigma_d). \tag{6}$$

To identify such discriminative model for our inference, a regularized (ridge regression) estimator in (7) is commonly used:

$$\hat{q} = \arg\min_q \sum_{i=1}^{h} (y_i - q^T w_i)^2 + 2\gamma||q||^2. \tag{7}$$

For the Normal model, we assume with the historical data are stored in

$$W_{\mathrm{mat}} = (w_1, w_2, \cdots, w_h), \quad Y_{\mathrm{mat}}^T = (y_1, y_2, \cdots, y_h), \tag{8}$$

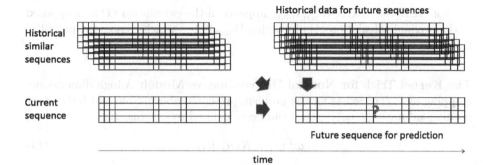

Fig. 7. Kernel ridge regression for prediction.

where the subscript h is the total number of chosen neighbor measurement sets, we can obtain a closed-form solution:

$$\hat{q} = (W_{\mathrm{mat}} W_{\mathrm{mat}}^{T} + 2\gamma I)^{-1} W_{mat} y. \tag{9}$$

The unknown hyper-parameter Σ_d has been absorbed into the penalty constant γ. Notice that due to the ridge regularization (since $\gamma > 0$), $W_{\mathrm{mat}} W_{\mathrm{mat}}^{T} + 2\gamma I \succeq 2\gamma I \succ 0$, so that this matrix is always invertible. Thus, the regularized estimator always exists. Once the hyper-parameter $\hat{q}$ is estimated, it can be used for the Bayesian inference to calculate the future measurement vector estimate $\hat{y}_{\mathrm{future}}^{B}$ as follows:

$$\hat{y}^{B} = \hat{q}^{T} w_{\mathrm{current}} = y_{\mathrm{current}}^{T} W_{mat}^{T} (W_{\mathrm{mat}} W_{\mathrm{mat}}^{T} + 2\gamma I)^{-1} w_{\mathrm{current}} = y_{\mathrm{current}}^{T} T. \tag{10}$$

Further by employing the Matrix Inversion Lemma $(A + BDC)^{-1} = A^{-1} - A^{-1}B(D^{-1} + CA^{-1}B)CA^{-1}$ to expand the inversion above, we can obtain the alternative form of T, which can further simplify the computation:

$$T = W_{mat}^{T} (W_{\mathrm{mat}} W_{\mathrm{mat}}^{T} + 2\gamma I)^{-1} w_{\mathrm{current}} = (W_{\mathrm{mat}}^{T} W_{\mathrm{mat}} + 2\gamma I)^{-1} W_{mat}^{T} w_{\mathrm{current}}, \tag{11}$$

where

$$W_{\mathrm{mat}}^{T} W_{\mathrm{mat}} = (w_1, w_2, \cdots, w_n)^{T} (w_1, w_2, \cdots, w_n)$$
$$= \begin{pmatrix} w_1^{T} w_1 & \ldots & w_1^{T} w_n \\ \vdots & \ddots & \vdots \\ w_n^{T} w_1 & \ldots & w_n^{T} w_n \end{pmatrix} \tag{12}$$

$$W_{\mathrm{mat}}^{T} w_{\mathrm{current}} = (w_1, w_2, \cdots, w_n)^{T} w_{current}$$
$$= \begin{pmatrix} w_1^{T} w_{current} \\ \vdots \\ w_n^{T} w_{current} \end{pmatrix}. \tag{13}$$

Note that the matrix $W_{mat}^T W_{mat}$ appears in the calculation (11), as opposed to the original calculation (11) involving $W_{mat} W_{mat}^T$, which creates the potential to reduce computation cost.

The Kernel Trick for Normal Discriminative Model: A high-dimensional mapping $\boldsymbol{w} = f(\boldsymbol{u})$ exists for the problem presented above, from which the inner product $\boldsymbol{w}_i^T \boldsymbol{w}_j = (f(\boldsymbol{u}_i))^T f(\boldsymbol{u}_j)$ can be calculated by a kernel $K(\cdot, \cdot)$, as below,

$$\boldsymbol{w}_i^T \boldsymbol{w}_j = K(\boldsymbol{u}_i, \boldsymbol{u}_j). \tag{14}$$

Be aware that kernel calculation uses only the (low-dimensional) $\boldsymbol{u}$'s, rather than the high-dimensional $\boldsymbol{w}$'s. Therefore, the computational complexity of calculating the inner products in (12) and (13) is low, even though $\dim(\boldsymbol{w}_{current})$ itself may represent a large number. This idea of using a cost-effective kernel calculation to implement a high-dimensional Normal model is called 'the kernel trick'. In this paper, we employ the following kernel forms as candidates.

- Homogeneous polynomial: $K(\boldsymbol{u}_i, \boldsymbol{u}_j) = (\boldsymbol{u}_i^T \boldsymbol{u}_j)^d$.
- Inhomogeneous polynomial: $K(\boldsymbol{u}_i, \boldsymbol{u}_j) = (1 + \boldsymbol{u}_i^T \boldsymbol{u}_j)^d$.
- Gaussian (Radial Basis function): $K(\boldsymbol{u}_i, \boldsymbol{u}_j) = \exp(-\mu ||\boldsymbol{u}_i^T \boldsymbol{u}_j||^2)$, $\mu > 0$.

Here d is chosen to be 2, because power flow model has a quadratic form of states. To choose a proper γ and kernel pair, we use the following training and validation phases.

- In the training phase, we apply a part of the historical data on different kernel functions and γs to calculate different Ts.
- In the validating phase, another part of historical data are used to validate the kernel function and γ.

Finally, the chosen $\hat{\boldsymbol{y}}^B$, computed from the validated T, is used in testing phase as an Nearest Neighbor (NN) Prediction for $\boldsymbol{y}_{future}$. As a result we obtain the prediction over the right bottom block in Fig. 7. Then this predicted block is combined with the left bottom block $\boldsymbol{z}_{current}^{seq}$, representing the current and short measurement sequence. Such combined sequence is used to find the most similar sequence in the historical database. If an event is associated with this historical sequence, the same event is declared for the new sequence.

4   Other PowerScope-Based Applications

Beside using data analysis to reduce searching time in the early event detection in this paper, one can also use the eigenvectors generated by PCA analysis to understand measurements within each cluster in Fig. 5b. For example, Fig. 8 shows typical daily power profiles. We list the possible physical explanation for the eight sub-figures as follows:

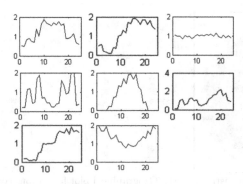

Fig. 8. Normalized active daily power flow/injection.

- Subplot 1: This daily power profile represents averaged users, who use power mostly in the day time. It can also represent oil, coal, gas generations, which catch up fast when the daily loads increase or decrease.
- Subplot 2 (first row, second column): This figure is similar to subplot 1. It could be another typical user/generator pattern.
- Subplot 3: The power consumption in this plot is flat, which indicates facilities consuming constant power, or a nuclear power plants, whose generation is usually fixed.
- Subplot 4: This daily profile looks like wind generators with a lot of randomness.
- Subplot 5: This daily profile has zero power generation at night, and monotonically increases in the morning and decreases in the evening. It behaves like a solar generators.
- Subplot 6: This profile represents special unknown users or generators.
- Subplot 7: This profile looks like a regulated industrial users, who is not allowed to use too much power during day time.
- Subplot 8: This profile represents smart users, who can reduce the power consumption in the noon due to economic/environmental reasons.

In the next two subsections, we illustrate the usage of this observation.

Percentage Analysis: By summarizing the total energy consumption within the power systems according to different eigenvalues, we can produce a pie chart of the percentage taken by different generation components within the power grid, i.e. Fig. 9a. This provides an important tool to understand how green a power grid is without conducting complicated network analysis, which can be plagued with outdated profiles or wrong information. Such a pie chart can be used for environmental, economic, and policy reasons.

By looking at the energy of singular values correlated to different daily power signals, we can see the rough percentages of different components in the grids. For example, Fig. 9a illustrates different components within the grid. There is 14 % green energy.

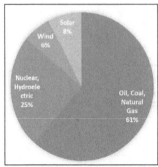

(a) Power component distribution analysis.

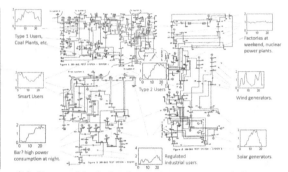

(b) Geographical plot for various components in power grid.

Fig. 9. Applications.

Geographical-Plot: A second application is the geographical plot of measurement data with different patterns. In Fig. 9b, we plot the daily active power flow, or power injection according to the correlation between the measurement and the eight typical daily load from SVD. This plot can show the system operator the geographical contributors of different major typical flows within the network purely based on the data.

5   Experiments

To verify the early event detection and identification ability of our proposed method, we carry out simulations in MATLAB using MATLAB Power System Simulation Package (MATPOWER) [13,14].

We use Matlab to generate historical data. To simulate the power system behavior which resembles real-world large power systems, we adopt the online load profile from New York ISO [15]. Specifically, we use the load data between February 2005 and September 2013 with a consistent data format. It has 11 online load profiles in New York ISO area, namely 'CAPITL', 'CENTRL', 'DUNWOD', 'GENESE', 'HUD VL', 'LONGIL', 'MHK VL', 'MILLWD', 'N.Y.C.', 'NORTH', and 'WEST'. The data is recorded every five minutes.

In order to obtain the 199 nonzero active load power consumptions in the IEEE 300 test case file, we employ a downsampling method to extend the 11 load profiles to 209 (11 × 19) profiles. Basically, instead of using the five minute interval from the online resource, we use a new interval of ninety five (5 × 19) min. Resulting from the down-sampling, we obtained 43, 379 valid historical load data for each load bus from February 2005 to December 2012. We also added random noise to mimic the stochastic behavior of system users. Generation is adaptively changed according to the total power balances, with Gaussian noise added to represent intermittence resources. The testing load profiles between July and October 2013 are generated using the same approach.

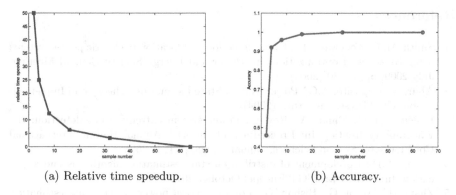

(a) Relative time speedup.   (b) Accuracy.

Fig. 10. Numerical results.

Second, we fit the load data into the case file, and change a topology connection with 10 % probability. Various event is added to the historical data. Next, we run an AC power flow to generate the true states of the power system, followed by creating 1073 true measurement sets with Gaussian noise and standard deviation around 0.015. Here, we assume that the measurement set includes (1) The power injection on each bus; (2) The transmission line power flow 'from' or 'to' each bus that it connects; (3) The direct voltage magnitude of each bus.

Figure 10 shows the numerical results. From it, we can see that when less sample is used, the speed up is large. For instance by using two samples, we can achieve 50 times speed up. The detection and identification accuracy is 46 %. Reversely, by using more samples, the accuracy is increasing while the speedup decreases. Notably, with only four samples, the detection and identification accuracy is more than 90 %.

Finally, a highlight feature of our proposed method is that no topology information is required for conducting it. Besides, our method, not only detects early events but also identifies it.

6 Conclusion

The challenges of smart gird are enormous as the power industry paradigm shifts from the traditionally complex physical model based monitoring architecture to data-driven model based resource management. In this paper, we propose a new historical data-driven early event detection method based on data analysis. In particular, we conduct Singular Value Decomposition and Independent Component Analysis for power grid data. Subsequently, we propose an early event detection and identification approach on the basis of nearest sub-sequence search and kernel regression model. This is based on the intuition that similar measurement sequences reflect similar power system events. Therefore, our data-driven method is a unique way for sustainable smart grid design to break the current model based monitoring architecture that requires both large complex modeling and computation overhead.

References

1. Smith, M.J., Wedeward, K.: Event detection and location in electric power systems using constrained optimization. In: Power and Energy Society General Meeting, July 2009, pp. 26–30 (2009)
2. Abur, A., Exposito, A.G.: Power System State Estimation: Theory and Implementation. CRC Press, New York (2004)
3. Lesieutre, B.C., Pinar, A., Roy, S.: Power system extreme event detection: the vulnerability frontier. In: Proceedings of the 41st Annual Hawaii International Conference on System Sciences, January 2008, p. 184 (2008)
4. Alvaro, L.D.: Development of distribution state estimation algorithms and application. In: IEEE PES ISGT Europe, October 2012
5. Zhang, J., Welch, G., Bishop, G.: Lodim: a novel power system state estimation method with dynamic measurement selection. In: IEEE Power and Energy Society General Meeting, July 2011
6. EATON: Power xpert meters 4000/6000/8000
7. Ilic, M.: Data-driven sustainable energy systems. In: The 8th Annual Carnegie Mellon Conference on the Electricity Industry, March 2012
8. Yang, C., Xie, L., Kumar, P.R.: Dimensionality reduction and early event detection using online synchrophasor data. In: Power and Energy Society General Meeting, July 2013, pp. 21–25 (2013)
9. Bishop, C.M.: Pattern Recognition and Machine Learning. Springer, New York (2006)
10. Strang, G.: Introduction to Linear Algebra (Sect. 6.7). Wellesley-Cambridge Press, Cambridge (1998)
11. Wikipedia: Independent component analysis, April 2014
12. Hyvarinen, A.: Fastica for matlab. http://research.ics.aalto.fi/ica/fastica/
13. Zimmerman, R.D., Murillo-Sanchez, C.E., Thomas, R.J.: Matpower's extensible optimal power flow architecture. In: IEEE Power and Energy Society General Meeting, July 2009, pp. 1–7 (2009)
14. Zimmerman, R.D., Murillo-Sanchez, C.E.: Matpower, a matlab power system simulation package, July 2010. http://www.pserc.cornell.edu/matpower/manual.pdf
15. NYISO: Load data profile, May 2012. http://www.nyiso.com

Machine Learning Techniques for Supporting Renewable Energy Generation and Integration: A Survey

Kasun S. Perera[1], Zeyar Aung[2(✉)], and Wei Lee Woon[2]

[1] Database Technology Group, Technische Universität Dresden, Dresden, Germany
Kasun.Perera@tu-dresden.de
[2] Institute Center for Smart and Sustainable Systems (iSmart),
Masdar Institute of Science and Technology, Abu Dhabi, UAE
{zaung,wwoon}@masdar.ac.ae

Abstract. The extraction of energy from renewable sources is rapidly growing. The current pace of technological development makes it commercially viable to harness energy from sun, wind, geothermal and many other renewable sources. Because of the negative effects on the environment and the economy, conventional energy sources like natural gas, crude oil and coal are coming under political and economic pressure. Thus, they require a better mix of energy sources with a higher percentage of renewable energy sources. Harnessing energy from renewable sources range from small scale (e.g., a single household) to large scale (e.g., power plants producing several MWs to a few GWs providing energy to an entire city). An inherent characteristic common to all renewable power plants is that power generation is dependent on environmental parameters and thus cannot be fully controlled or planned for in advance. In a power grid, it is necessary to predict the amount of power that will be generated in the future, including those from the renewable sources, as fluctuations in capacity and/or quality can have negative impacts on the physical health of the entire grid as well as the quality of life of its users. As renewable power plants continue to expand, it will also be necessary to determine their optimal sizes, locations and configurations. In addition, management of the smart grid, in which the renewable energy plants are integrated, is also a challenging problem. In this paper we provide a survey on different machine learning techniques used to address the above issues related to renewable energy generation and integration.

Keywords: Renewable energy · Smart grids · Machine learning

1 Introduction

The world is faced with a number of challenges related to energy sustainability and security. If not promptly addressed, these can lead to economic and political instability. The depletion of fossil fuel reserves as well as the environmental

© Springer International Publishing Switzerland 2014
W.L. Woon et al. (Eds.): DARE 2014, LNAI 8817, pp. 81–96, 2014.
DOI: 10.1007/978-3-319-13290-7_7

impact of burning these fuels have led to increased interest in developing alternative and more sustainable energy sources. Renewable energy resources like solar photovoltaic (PV), solar thermal (a.k.a. concentrated solar power, CSP), geothermal, tidal waves, wind power, and biomass have been growing rapidly in energy market [1]. Many countries and companies are seeking to diversify their energy mix by increasing the share of renewables.

In conventional energy generation process, energy production depends on the energy demand from the users, and the stability of the power grid relies on the equilibrium of energy demand and supply. When the energy demand surpasses the energy supply, it destabilizes the power grid and results in power quality degradation and/or blackouts in some parts of the grid. When the demand is lower than the supply, energy is lost incurring high unnecessary costs due to wastage. Producing the right amount of energy at the right time is crucial both for the smooth running of the grid and for higher economic benefits. To maintain this stability, much research has focused on energy supply and demand forecasting to predict the amount of energy that will be required. This will then ensure that there will be sufficient capacity to meet these requirements, but also that excess capacity and hence energy wasted will be minimized.

Renewable energy resources like solar light, solar heat and wind are highly variable and the resulting fluctuations in the generation capacity can cause instability in the power grid. This is because the energy/power output of these plants is defined by the environmental factors such as wind speed, the intensity of solar radiation, cloud cover and other factors. Another important limitation of renewable energy power plants is that they are subject to marked daily and annual cycles (e.g., solar energy is only available during the day). Thus, it is necessary to generate power when resources are available and store it for later use while using a certain portion of the generated power at the same time. Wind and solar PV energy is expensive to store, thus careful management of energy generation is needed. When the generation capacity of natural resources are insufficient to meet demand, conventional sources such as gas power plants are typically used to cover the electricity shortfall.

The above-mentioned challenges have motivated the use of machine learning techniques to support better management of energy generation and consumption. Different machine learning techniques are used in different stages of a renewable energy-integrated power grid, depending on the requirements and the characteristics of the problem. For a power grid with renewable energy sources contributing a considerable proportion of energy supply, it is necessary to forecast both short and medium term demand. This would facilitate the formulation of well informed energy policies, for example by helping to determine important parameters such as the appropriate spinning reserve levels and storage requirements. On the other hand, it is also necessary to forecast the energy output from renewable energy power plants themselves, since the energy output from these power plants depends on many environmental factors that cannot be controlled. This in turn necessitates the prediction of these environmental factors such as wind speed, direction and solar radiation in the region of the power plant.

Another important use for machine learning techniques in the context of renewable energy is in determining the optimal location, size and configuration of renewable power plants. These parameters are dependent on many factors such as proximity to population centers, local climatic fluctuations, terrain, availability and costs of logistics and other facilities and many others. Yet another area for the application of machine learning methods is in the overall operations and management of the smart grid, i.e. issues such as fault detection, control and so on.

Figure 1 depicts possible areas where we can use machine learning techniques for performance improvements and better management of renewable energy. The right side of the figure depicts consumers and prosumers (who consume energy from the grid as well as produce small-scale renewable energy and feed the excessive energy to the grid). The left side depicts large-scale renewable energy producers. Conventional power plants are still involved in the grid in order to balance of demand and supply and to ensure adequate power quality (Table 1).

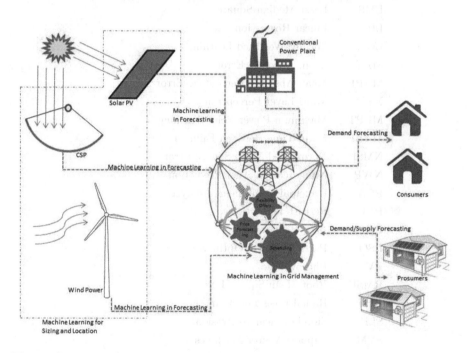

Fig. 1. Overview of power grid with integrated renewable sources and its usage of machine learning techniques in different steps.

This paper will summarize and compare the machine learning techniques that have been or can be used not just in the generation of renewable energy but also in the integration of these resources into existing power grids.

The rest of the paper is organized as follows. Section 2 describes the machine learning techniques in power output prediction from different renewable sources.

Table 1. List of acronyms.

Acronym	Meaning
ANN	Artificial Neural Network
AR	Additive Regression
ARIMA	Auto-Regressive Integrated Moving Average
ARMA	Auto-Regressive Moving Average
CART	Classification and Regression Trees
CSP	Concentrated Solar Power
DEA	Data Envelopment Analysis
FFT	Fast Fourier Transformation
GA	Genetic Algorithm
kNN	k-Nearest Neighbor
LLP	Loss of Load Probability
LMS	Least Median Square
LR	Linear Regression
LWL	Locally Weighted Learning
MAE	Mean Absolute Error
MAPE	Mean Absolute Percentage Error
MLP	Multi-Layer Perceptron
MPPT	Maximum Power Point Tracking
MTBF	Mean Time Between Failures
NMSE	Normalized Mean Square Error
NWP	Numerical Weather Prediction
PCA	Principal Component Analysis
P&O	Perturb and Observe
PR	Pace Regression
PSO	Particle Swarm Optimization
PV	Photovoltaic
RMSE	Root Mean Square Error
RBF	Radial Basis Function
SLR	Simple Linear Regression
SVM	Support Vector Machines

Section 3 discusses the techniques used in optimizing location, sizing, and configurations of renewable sources. Section 4 covers the methods in overall operations and management of a mixed-source smart grid with renewable energy playing a significant role. Finally, Sect. 5 concludes the paper.

2   Forecasting Renewable Energy Generation

Forecasting power output from a renewable energy power plant is crucial as this depends on many non-human-controllable factors such as environmental parameters. Depending on the energy source it uses, the power plant exhibits certain characteristics that enables the use of machine learning techniques for prediction purposes. In this section we will review the different machine learning techniques used in different types of power plants including wind farms, solar farms, and hydro power.

2.1   Wind Power Generation

Wind power generation depends on many characteristics and the power output from a wind turbine can be calculated using the Eq. 1. Here A stands for area that is covered by the wind turbine blades (a circle with radius r), ρ is for air density, V is wind speed and C_p for efficiency factor usually imposed by the manufacturer.

$$P = \frac{A\rho V^3 C_p}{2} \tag{1}$$

In this equation wind speed is a significant factor as the power output is proportional to the wind speed. It also observed that there is a cutoff speed where the power output is steady after that speed (so as to ensure the safety of the turbine). Other factors such as humidity and temperature also affect the density of the air, which in turn affects the power generation. Thus, it is necessary to forecast these factors and ultimately the final power output in a wind farm. Many methods have been proposed for forecasting power generation in wind farms. Brief descriptions and reviews on them are given below.

In [2], Lei et al. presented physical and statistical-based models as two main categories of forecasting models. The physical models are more suitable for long term forecasting whereas the statistical models are used for short and medium term forecasting. Our interest lies in the statistical models as they are more closely associated with machine learning techniques.

Auto-regressive moving average (ARMA) and auto-regressive integrated moving average (ARIMA) models are presented in [3] for wind speed forecasting and then wind power forecasting by analyzing the time-series data. The authors start with the well known ARMA model and then apply ordered differential transformation to the model to get the ARIMA model. The ARMA model is a combination of AR model and MA model on the same time series data.

A Kalman filter model using the wind speed as the state variable is used in [4]. The authors suggested that this model is suitable for online forecasting of wind speed and generated power. Online forecasting of power generation is important as it can provide the most recent and updated future forecasting which then can be used for power grid management.

Comparison of the ARIMA and ANN models for wind speed forecasting in Oaxaca region in Mexico is presented in [5] by Sfetsos. Their analysis showed that seasonal ARIMA model outperformed ANN model for more accurate forecasting, but when the number of training vectors were increased for ANN model its accuracy could be improved. Using the previous ten-minute data for training, Sfetsos also presented a model [6] using ANN for wind time-series forecasting. Subsequent predictions are averaged to obtain the mean hourly wind speed and then to determine the power generation from the wind turbine.

Recurrent multi-layer perceptron (MLP) model, a variant of ANN, was proposed in [7], which employs Kalman filter based back-propagation network. The proposed model performs well in long term power generation prediction than in short term prediction.

In Mohandes et al. [8] an SVM using Gaussian kernels was used to predict the wind speed. The proposed method performed better than the MLP in terms the root mean square error (RMSE) on 12 years of wind data from Medina city, Saudi Arabia.

Fuzzy models are another way of using machine learning for prediction. In [9], Damousis et al. used a fuzzy model with spatial correlation method for wind power generation prediction. The proposed model performs well on wind turbines installed in a flat terrain, but performs poorly with respect to those installed in a deteriorated terrain. This might be due to variation of the wind speed with respect to height of the tower from the ground level as well as quality differences in the air.

Numerical weather prediction (NWP) models [10] were also used for wind forecasting and subsequently power generation prediction in many research works. In this approach, selecting an accurate NWP model is crucial as the accuracy well depends on the initial NWP model. In order to mitigate the effect of single NWP model, an ensemble method was proposed in [11]. The ensemble model allows to use same NWP with different parameters such as different initial conditions, physical parameterization of the sub-grid system or different data assimilation systems. It also can employ completely different NWP models to obtain the final ensemble learner.

Jursa and Rohrig [12] presented a mixed model using k-nearest neighbor (kNN) and ANN approaches. Their optimization model produces the results with 10.75 % improvement over the benchmark model (persistence method) used with respect to RMSE. Jursa [13] also proposed the use of a variety of machine learning models. In that work, wind power forecasts were determined at 10 wind farms and compared to the NWP data at each wind farm using classical ANNs, mixture of experts, SVM and kNN with particle swarm optimization (PSO). The main conclusion was that combining several models for day-ahead forecasts produces better results.

Foley et al. [14] provides a good survey of methods for wind power generation forecasting. It listed SVM, MLP, ANN, regression trees, and random forest as the widely-used machine learning methods in the context of wind power.

2.2 Solar Energy Generation

Solar photovoltaic (PV) usage ranges from the single household level to large solar PV plants with capacities of 1–100 MW. As solar PV have been used in small domestic level for a long time, a number of research works for performance estimation for PV using machine learning techniques have been conducted in the past years.

Thermo siphon solar heater is a way of using renewable energy to get hot water for domestic usage. Kalogirou et al. [15] conducted performance prediction for these devices using ANN. The performance was measured in terms of the useful energy extracted and of the stored water temperature rise. The ANN was trained using the performance data for four types of systems, all employing the same collector panel under varying weather conditions. The output of ANN is the useful energy extracted from the system and the water temperature rise. Seven input units, 24 hidden neurons and 2 neurons as output comprises the network model with Sigmod as the transfer function.

A site specific prediction model for solar power generation based on weather parameters was proposed in [16], in which Sharma et al. used different machine learning techniques. Multiple regression techniques including least-square SVM using multiple kernel functions were used in the comparison with other models. Experimental results showed that the SVM model outperformed the others with up to 27 % more accuracy. Linear least-square regression model was also used for prediction with 7 weather parameters and results indicate $165\,W/m^2$ and $130\,W/m^2$ RMSE for validation and prediction sets respectively. For the SVM-based model they tired linear, polynomial and RBF kernels, and chose the RBF kernel for the final SVM model (as the first two did not perform well). Further improvement to the model was made by using principal component analysis (PCA), thus by selecting the first 4 features from the ranked output features from PCA.

A hybrid intelligent predictor for 6 h ahead solar power prediction was proposed in [17]. The system used an ensemble method with 10 widely-used regression models namely linear regression (LR), radial basis function (RBF), SVM, MLP, pace regression (PR), simple linear regression (SLR), least median square (LMS), additive regression (AR), locally weighted learning (LWL) and IBk (an implementation of kNN). Their results showed that, with respect to MAE and MAPE, the top most accurately performing regression models are LMS, MLP, and SVM.

Diagne et al. [18] recently provided a survey of solar energy forecasting methods covering various physical and statistical/machine learning techniques.

2.3 Hydro Power Generation

Hydro power is the most widely used and one of the most established renewable energy sources. Due to its characteristics and economic viability, many third world countries depend on extracting energy from their available water sources. As hyrdo power uses running water or stored water sources which depend on the rainfall in the region, it is obviously affected by non-human controllable weather parameters which need to be forecasted for better planning and management.

Recurrent ANNs [19–21] as well as SVMs have been widely used for rainfall prediction. In [22], Hong presented the use of a combination of recurrent ANN and SVM to forecast rainfall depth values. Moreover, chaotic PSO algorithm was employed to choose the parameters for the SVM model. With the 129 data points provided to the model, the resulting performance of the model in terms of the normalized mean square error (NMSE) values were 1.1592, 0.4028 and 0.3752 for training, validation, and testing sets respectively.

An ensemble learning model for hydropower energy consumption forecasting was presented in [23], where Wang et al. used a seasonal decomposition-based least-squares SVM mechanism. The original time series data was decomposed into regional factors (which demonstrate seasonal effects/trends) and irregular components, and then all of them were used for least-square SVM analysis. This least-square SVM model was used to predict the three main components known as trend cycle, seasonal factor, and irregular component which were in turn fed into another least-square SVM to combine the prediction values. The authors stated that the model outperformed the other benchmark models by providing accurate results when seasonal effects and irregularities were presented in the input time-series.

Lansberry and Wozniak [24] used a genetic algorithm (GA) to support optimal governor tuning in hydro power plants. The authors investigated the GA as one possible means for adaptively optimizing the gains of proportional-plus-integral governors. This tuning methodology was adaptive towards changing plant parameters-conduit time constant and load self-regulation.

Djukanovic et al. [25–27] presented an ANN-based coordinated control for both exciter and governor for low head hydropower plants. Their design was based on self-organization and the predictive estimation capabilities of ANN implemented through the cluster-wise segmented associative memory scheme [25].

3   Determining Plant Location, Size, and Configuration

Unlike natural gas, diesel or coal fired plants, renewable energy power plants require a huge area for their operation. For example, Shams-1, which is the biggest CSP power plant in the world opened recently in Abu Dhabi, UAE, occupies an area of $2\,km^2$ and generates $100\,MW$ of electricity. A conventional power plant of similar capacity only takes a few square meters space. Thus, it is necessary to analyze the required size of the renewable energy power plant with respect to the energy requirements. Power plants like solar PV and CSP also exhibits special requirements of location selection and orientation selection as solar panels need to be faced to solar irradiation to absorb the optimal energy. Thus, machine learning techniques play a crucial role in assisting these decision making steps.

Conventional methods for sizing PV plants have generally been used for locations where the required weather data (irradiation, temperature, humidity, clearness index, wind speed, etc.) is available and so is the other information concerning the site where the PV plant is to be built. However, these methods could not be used for sizing PV systems in remote areas where the required data are not readily available, and thus machine learning techniques are needed to be employed for estimation purposes.

Mellit et al. [28] developed an ANN model for estimating sizing parameters of stand-alone PV systems. In this model, the inputs are the latitude and longitude of the site, while the outputs are two hybrid-sizing parameters (f, u). These parameters are determined by simple regression of loss of load probability (LLP) as shown in Eq. 2.

$$f = f_1 + f_2 \log(LLP) \quad and \quad u = e^{(u_1 + u_2 \cdot LLP)} \tag{2}$$

These parameters allow the designers of PV systems to determine the number of solar PV modules and the storage capacity of the batteries necessary to satisfy demand. In the proposed model, the relative error with respect to actual data does not exceed 6 %, thus providing accurate predictions. In addition, radial basis function network has been used for identification of the sizing parameters of the PV system. Their model, depicted in Fig. 2, has been evaluated on 16 different sites and experimental results indicated that prediction error ranges from 3.75 %–5.95 % with respect to the sizing parameters f and u.

Seeling-Hochmuth [29] presented research into the optimization of PV-hybrid energy systems. The proposed method optimizes the configuration of the system and the control strategy by means of a GA. The control of the system is coded as a vector whose components are five decision variables for every hour of the year.

Senjyua et al. [30] also developed an optimal configuration of power generating systems in isolated islands with renewable energy using a GA. The hybrid power generation system consisted of diesel generators, wind turbine generators, PV system and batteries. The proposed methodology can be used to determine the optimum number of solar array panels, wind turbine generators, and battery configurations. The authors argued that by using the proposed method, operation cost can be reduced by about 10 % in comparison with using diesel generators only.

Similarly, Yokoyama et al. [31] proposed a multi-objective optimal unit sizing of hybrid power generation systems utilizing PV and wind energy.

Hernadeza et al. [32] presented a GA-based approach to determining the optimal allocation and sizing of PV grid connected systems in feeders that provides the best overall impact on the feeder. The optimal solution is reached by a multi-objective optimization approach. According to the authors, the results obtained with the proposed methodology for feeders performed well when compared with the results found in the literature.

A flexible neuro-fuzzy approach for location optimization of solar plants with possible complexity and uncertainty was described in [33]. The flexible approach was composed of ANN and fuzzy data envelopment analysis (DEA). In the model, first fuzzy DEA was validated by ordinary DEA, and then it was used for ranking of solar plant units and the best α-cut was selected based on the test of normality. Several ANN models are developed using MLP and the network with minimum MAPE was selected for the final model building.

Maximum power point tracking (MPPT) in solar PV is essential as it helps to extract the maximum energy in a given time period. Rotational non-static PV

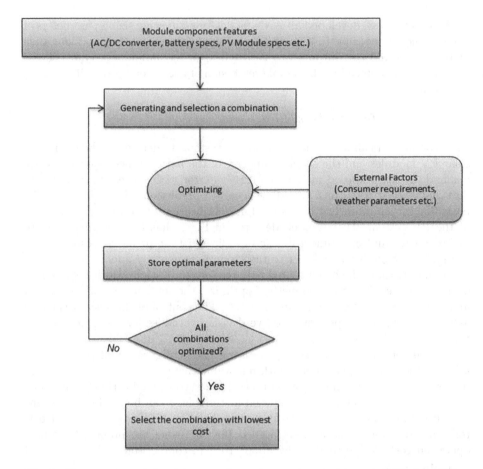

Fig. 2. The overview of the sizing, configuration, and optimizing of a PV plant [28].

panels are employed with intelligent mechanisms for sun tracking. An intelligent MPPT was outlined in [34], which used fuzzy logic approach with the perturb and observe (P&O) algorithm. The subjective fuzzy model of the system was designed based on prior expert knowledge of the system. The Fuzzy logic controller was divided into four sections: fuzzification, rule-base, inference and defuzzification.

4   Managing Renewable Energy-Integrated Smart Grid

As rapid advancements in the power grid continue to make it smarter, its users/stakeholders expect more efficient and effective operation and management of the grid. Since more and more stakeholders take part in the power grid, managing such a big network becomes harder. Thus, intelligent techniques are required to cater the better management of the smart grid. In this section we will outline some problems the power grids are facing, namely supply/damand balancing, grid operations and management, grid's data management, and the proposed machine learning solutions to them. In addition, we will briefly describe a promising approach for the grid's data management problem.

4.1   Balancing Supply and Demand

When a power grid is integrated with renewable sources, it is even more important to accurately forecast energy generation as well as energy consumption. Fluctuations and intermittent behavior of solar and wind power plants imposes vulnerabilities to the power grid, thus by destabilizing the grid. Therefore, in order to maintain the stability of the grid, it is necessary to connect to the conventional power generation in time, disconnect malfunctioning wind power plants/turbines, or use smoothing techniques for the solar PV plants and grid connection. To identify those factors affecting the grid's stability and to ensure its good management, various machine learning techniques were employed.

The MIRABEL [35] system offers forecasting models which target flexibilities in energy supply and demand, thus helping to manage the production and consumption in the smart grid with renewable energy plants. The forecasting model can also efficiently process new energy measurements to detect changes in the upcoming energy production or consumption and to enable the rescheduling of flex-offers if necessary. The model uses a combination of widely adopted algorithms like SVM and ensemble learners. In order to better manage the demand and supply depending on the time domain, it employs different models for different time scales.

In a near future, the smart grid will consist of many individual autonomous units such as smart households, smart offices or smart vehicles. These users on the demand side pose a varying demand on the grid as their requirements are changing over the time and their life styles. Moreover, the demand is also affected by pricing regulations of the grid, as the smart grid employ deregulated pricing mechanisms at many levels. This deregulated market offers a flexibly to the users, thus allowing them to bid for energy that they need. Forecasting those flexibility offers is crucial in the smart grid systems today. Barbato et al. [36] and Reinhardt et al. [37] forecasted the energy demand through meter readings from households. This provides detecting the flexible energy usage from the connected autonomous users in the demand side. Kaulakiene et al. extended that idea in [38] by suggesting methods to extract the flexibilities from the electricity time series.

On the energy supply side, since there are different stakeholders with different characteristics, predicting energy supply can be a very challenging task. Use of multiple models to predict the energy supply is a common approach among the users. This impose another challenge as there is no systematic method to select which models to use when necessary. Ulbricht et al. [39] presented a systematical optimized strategy to select suitable models from a model pool to use for solar energy supply forecasting.

4.2   Grid Operations and Management

For the operations and management of the grid itself, an overview of machine learning techniques used in New York City power grid was provided in [40] by Rudin et al. The system consisted of many different models for forecasting in different levels of the entire power grid. These models can be used directly by power companies to assist with prioritization of maintenance and repair work. Specialized versions of the proposed process are used to produce (1) feeder failure rankings, (2) cable, joint, terminator, and transformer rankings, (3) feeder mean time between failures (MTBF) estimates, and (4) manhole events vulnerability rankings. In their model, the authors used approximately 300 features generated from the time series data and associated parameters. SVM, classification and regression trees (CART), ensemble learning techniques such as random forests, and statistical methods were used for model building.

4.3   Grid Data Management

As smart grid deployments continue to expand via the addition of more users, it often requires information to be exchanged amongst different stakeholders. Many users generated frequent data that need to be shared among interesting parties to help make decisions for better management of the smart grid. So, it is required to employ a efficient and effective methods to share the smart grid's data.

Compression is a widely used technique to help data exchange when it has to deal with large quantities of data. Louie and Miguel [41] presented a lossless compression of wind plant data by using characteristics related to the wind plants. They suggests two methods to use with grid based wind plants and un-ordered wind plants. The authors claimed the superiority of their methods having 50 % more compression than state-of-the-art methods and managed to achieve ∼1.8 times compression rate. In simple terms 1.8 compression rate means 1.8 GB of data compressed to 1 GB data, thus, user only has to work on 1 GB data instead of 1.8 GB of data.

Reeves et al. [42] described a model-based compression of massive time series data. The authors presented a method employing fast Fourier transformation (FFT), filtered spikes and random signal projection to represent the original time series data. The method achieved a data reduction rate of 91.5 %. The method was a lossy compression but still preserved the important information.

Here, we suggest that a similar model-based data representation method be used for smart grid data as it can potentially provide many advantages.

1. Model-based representation provides low memory footprint while maintaining the same required information.
2. It also provides efficient method to exchange information, since it does not need to exchange raw data, but can exchange the representation model.
3. Models provide efficient query processing when compared to query processing on raw data.

We envision that the smart grid will greatly benefit from model-based data representation. Further analysis and results will be included in the upcoming research papers.

5 Conclusion

Due to the depletion of conventional energy sources like natural gas, crude oil and coal and to mitigate the environmental effects of the burning of fossil fuels, governments and companies are focusing increasingly on developing renewable energy sources. Hydro power is a good example of a renewable source that has been successfully used for many decades. Wind and solar are also promising renewable sources that have experienced a fast pace of growth in the recent years. An inherent feature of these resources is that the energy production capacity is not fully controllable or even predictable, thus necessitating the use of proper forecasting and management techniques to ensure smooth integration with the power grid. A smart power grid that incorporates renewable energy sources needs to be constantly monitored and need to have the forecasting ability to predicting sudden changes in the power supply and demand. The studies reviewed in this paper analyze the different machine learning techniques used for supporting the generation of renewable energy and more importantly their integration into the power grid. It is very difficult to generalize the machine learning models for each and every aspect of renewable energy generation and integration into the grid, but a strong coordination is necessary among the different prediction and decision making models to better enhance the grid's overall efficiency and effectiveness.

In addition, machine learning techniques have been successfully used in the planning of renewable energy plants based on available data with reasonable accuracy. Published literature on location, sizing, and configuration of wind and PV systems based on machine learning techniques underline their popularity, particularly in isolated areas. This shows the potential of machine learning as a design tool in strategic planning and policy making for renewable energy.

References

1. United States Department of Energy: 2012 renewable energy data book (2012). http://www.nrel.gov/docs/fy14osti/60197.pdf
2. Lei, M., Shiyan, L., Chuanwen, J., Hongling, L., Yan, Z.: A review on the forecasting of wind speed and generated power. Renew. Sustain. Energ. Rev. **13**, 915–920 (2009)
3. Guoyang, W., Yang, X., Shasha, W.: Discussion about short-term forecast of wind speed on wind farm. Jilin Electr. Power **181**, 21–24 (2005)
4. Ding, M., Zhang, L.J., Wu, Y.C.: Wind speed forecast model for wind farms based on time series analysis. Electr. Power Autom. Equipment **25**, 32–34 (2005)
5. Sfetsos, A.: A comparison of various forecasting techniques applied to mean hourly wind speed time series. Renew. Energ. **21**, 23–35 (2000)

6. Sfetsos, A.: A novel approach for the forecasting of mean hourly wind speed time series. Renew. Energ. **27**, 163–174 (2002)
7. Li, S.: Wind power prediction using recurrent multilayer perceptron neural networks. In: Proceedings of the 2003 IEEE Power Engineering Society General Meeting, vol. 4, pp. 2325–2330 (2003)
8. Mohandes, M.A., Halawani, T.O., Rehman, S., Hussain, A.A.: Support vector machines for wind speed prediction. Renew. Energ. **29**, 939–947 (2004)
9. Damousis, I.G., Alexiadis, M.C., Theocharis, J.B., Dokopoulos, P.S.: A fuzzy model for wind speed prediction and power generation in wind parks using spatial correlation. IEEE Trans. Energ. Convers. **19**, 352–361 (2004)
10. Giebel, G., Badger, J., Landberg, L., et al.: Wind power prediction ensembles. Technical report 1527, Ris National Laboratory, Denmark (2005)
11. Al-Yahyai, S., Charabi, Y., Al-Badi, A., Gastli, A.: Nested ensemble NWP approach for wind energy assessment. Renew. Energ. **37**, 150–160 (2012)
12. Jursa, R., Rohrig, K.: Short-term wind power forecasting using evolutionary algorithms for the automated specification of artificial intelligence models. Int. J. Forecast. **24**, 694–709 (2008)
13. Jursa, R.: Wind power prediction with different artificial intelligence models. In: Proceedings of the 2007 European Wind Energy Conference and Exhibition (EWEC), pp. 1–10 (2007)
14. Foley, A.M., Leahy, P.G., Marvuglia, A., McKeogh, E.J.: Current methods and advances in forecasting of wind power generation. Renew. Energ. **37**, 1–8 (2012)
15. Kalogirou, S.A., Panteliou, S., Dentsoras, A.: Artificial neural networks used for the performance prediction of a thermosiphon solar water heater. Renew. Energ. **18**, 87–99 (1999)
16. Sharma, N., Sharma, P., Irwin, D., Shenoy, P.: Predicting solar generation from weather forecasts using machine learning. In: Proceedings of the 2011 IEEE International Conference on Smart Grid Communications (SmartGridComm), pp. 528–533 (2011)
17. Hossain, M.R., Amanullah, M.T.O., Shawkat Ali, A.B.M.: Hybrid prediction method of solar power using different computational intelligence algorithms. In: Proceedings of the 22nd Australasian Universities Power Engineering Conference (AUPEC), pp. 1–6 (2012)
18. Diagne, M., David, M., Lauret, P., Boland, J., Schmutz, N.: Review of solar irradiance forecasting methods and a proposition for small-scale insular grids. Renew. Sustain. Energ. Rev. **27**, 65–76 (2013)
19. Kechriotis, G., Zervas, E., Manolakos, E.S.: Using recurrent neural networks for adaptive communication channel equalization. IEEE Trans. Neural Netw. **5**, 267–278 (1994)
20. Jordan, M.I.: Attractor dynamics and parallelism in a connectionist sequential machine. In: Proceeding of 8th Annual Conference of the Cognitive Science Society (CogSci), pp. 531–546 (1987)
21. Williams, R., Zipser, D.: A learning algorithm for continually running fully recurrent neural networks. Neural Comput. **1**, 270–280 (1989)
22. Hong, W.C.: Rainfall forecasting by technological machine learning models. Appl. Math. Comput. **200**, 41–57 (2008)
23. Wang, S., Tang, L., Yu, L.: SD-LSSVR-based decomposition-and-ensemble methodology with application to hydropower consumption forecasting. In: Proceedings of the 4th International Joint Conference on Computational Sciences and Optimization (CSO), pp. 603–607 (2011)

24. Lansberry, J.E., Wozniak, L.: Optimal hydro generator governor tuning with a genetic algorithm. IEEE Trans. Energ. Convers. **7**, 623–630 (1992)
25. Djukanovic, M., Novicevic, M., Dobrijevic, D.J., et al.: Neural-net based coordinated stabilizing control for the exciter and governor loops of low head hydropower plants. IEEE Trans. Energ. Convers. **10**, 760–767 (1995)
26. Djukanovic, M.B., Calovic, M.S., Vesovic, B.V., Sobajic, D.J.: Neuro-fuzzy controller of low head power plants using adaptive-network based fuzzy inference system. IEEE Trans. Energ. Convers. **12**, 375–381 (1997)
27. Djukanovic, M.B., Calovic, M.S., Vesovic, B.V., Sobajic, D.J.: Coordinated-stabilizing control for the exciter and governor loops using fuzzy set theory and neural nets. Int. J. Electr. Power Energ. Syst. **19**, 489–499 (1997)
28. Mellit, A., Benghanem, M., Hadj Arab, A., Guessoum, A.: Modeling of sizing the photovoltaic system parameters using artificial neural network. In: Proceedings of the 2003 IEEE International Conference on Control Application (CCA), pp. 353–357 (2003)
29. Seeling-Hochmuth, G.C.: Optimisation of hybrid energy systems sizing and operation control. Ph.D. thesis, University of Kassel, Germany (1998)
30. Senjyua, T., Hayashia, D., Yonaa, A., Urasakia, N., Funabashib, T.: Optimal configuration of power generating systems in isolated island with renewable energy. Renew. Energ. **32**, 1917–1933 (2007)
31. Yokoyama, R., Ito, K., Yuasa, Y.: Multiobjective optimal unit sizing of hybrid power generation systems utilizing photovoltaic and wind energy. J. Sol. Energ. Eng. **116**, 167–173 (1994)
32. Hernádeza, J.C., Medinaa, A., Juradob, F.: Optimal allocation and sizing for profitability and voltage enhancement of PV systems on feeders. Renew. Energ. **32**, 1768–1789 (2007)
33. Azadeh, A., Sheikhalishahi, M., Asadzadeh, S.M.: A flexible neural network-fuzzy data envelopment analysis approach for location optimization of solar plants with uncertainty and complexity. Renew. Energ. **36**, 3394–3401 (2011)
34. Revankar, P.S., Thosar, A.G., Gandhare, W.Z.: Maximum power point tracking for PV systems using MATLAB/SIMULINK. In: Proceedings of the 2nd International Conference on Machine Learning and Computing (ICMLC), pp. 8–11 (2010)
35. Boehm, M., Dannecker, L., Doms, A., et al.: Data management in the MIRABEL smart grid system. In: Proceedings of the 2012 Joint EDBT/ICDT Workshops, pp. 95–102 (2012)
36. Barbato, A., Capone, A., Rodolfi, M., Tagliaferri, D.: Forecasting the usage of household appliances through power meter sensors for demand management in the smart grid. In: Proceedings of the 2011 IEEE International Conference on Smart Grid Communications (SmartGridComm), pp. 404–409 (2011)
37. Reinhardt, A., Christin, D., Kanhere, S.S.: Predicting the power consumption of electric appliances through time series pattern matching. In: Proceedings of the 5th ACM Workshop on Embedded Systems for Energy-Efficient Buildings (BuildSys), pp. 30:1–30:2 (2013)
38. Kaulakienė, D., Šikšnys, L., Pitarch, Y.: Towards the automated extraction of flexibilities from electricity time series. In: Proceedings of the 2013 Joint EDBT/ICDT Workshops, pp. 267–272 (2013)
39. Ulbricht, R., Fischer, U., Lehner, W., Donker, H.: First steps towards a systematical optimized strategy for solar energy supply forecasting. In: Proceedings of the 2013 ECML/PKDD International Workshop on Data Analytics for Renewable Energy Integration (DARE), pp. 14–25 (2013)

40. Rudin, C., Waltz, D., Anderson, R., et al.: Machine learning for the New York City power grid. IEEE Trans. Pattern Anal. Mach. Intell. **34**, 328–345 (2012)
41. Louie, H., Miguel, A.: Lossless compression of wind plant data. IEEE Trans. Sustain. Energ. **3**, 598–606 (2012)
42. Reeves, G., Liu, J., Nath, S., Zhao, F.: Managing massive time series streams with multi-scale compressed trickles. Proc. VLDB Endowment **2**, 97–108 (2009)

A Framework for Data Mining in Wind Power Time Series

Oliver Kramer[1]($\boxtimes$), Fabian Gieseke[2], Justin Heinermann[1], Jendrik Poloczek[1], and Nils André Treiber[1]

[1] Computational Intelligence Group, Department of Computer Science,
University of Oldenburg, Oldenburg, Germany
{oliver.kramer,justin.heinermann,jendrik.poloczek,
nils.andre.treiber}@uni-oldenburg.de
[2] Image Group, Department of Computer Science, University of Copenhagen,
Copenhagen, Denmark
fabian.gieseke@di.ku.dk

Abstract. Wind energy is playing an increasingly important part for ecologically friendly power supply. The fast growing infrastructure of wind turbines can be seen as large sensor system that screens the wind energy at a high temporal and spatial resolution. The resulting databases consist of huge amounts of wind energy time series data that can be used for prediction, controlling, and planning purposes. In this work, we describe WindML, a Python-based framework for wind energy related machine learning approaches. The main objective of WindML is the continuous development of tools that address important challenges induced by the growing wind energy information infrastructures. Various examples that demonstrate typical use cases are introduced and related research questions are discussed. The different modules of WindML reach from standard machine learning algorithms to advanced techniques for handling missing data and monitoring high-dimensional time series.

1 Introduction

Wind is an important renewable energy resource. With the growing infrastructure of wind turbines and the availability of time series data given at a high spatial and temporal resolution, the application of machine learning methods comes into play. In most cases, the use of such tools affords multiple preprocessing steps to extract reasonable patterns from the raw wind time series data. To facilitate these steps, we propose a new machine learning and data mining framework called WINDML.[1] It offers machine learning methods adapted to wind energy time series tasks including standard ones such as classification, regression, clustering, and dimensionality reduction, which in turn can be useful for prediction, controlling, and planning purposes in the context of modern energy systems. Similar frameworks for other domains have been introduced over the past years

---

[1] The source code is publicly available on http://www.windml.org.

© Springer International Publishing Switzerland 2014
W.L. Woon et al. (Eds.): DARE 2014, LNAI 8817, pp. 97–107, 2014.
DOI: 10.1007/978-3-319-13290-7_8

(e.g., ASTROML [16] for astronomy), and the WINDML framework follows this line of research for the emerging field of modern wind energy infrastructures. It is based on PYTHON and builds upon various well-known packages including NUMPY [17], SCIPY [5], MATPLOTLIB [4], and SCIKIT-LEARN [9]. It is easily expansible w.r.t. the variety of methods and of wind time series databases. Further, WINDML is tailored to the specific needs of both the development of new machine learning methods and the application of tools for educational purposes.

This work is organized as follows. In Sect. 2, general properties of the WINDML framework and the employed databases will be introduced. Section 3 provides a description of some exemplary use cases such as wind energy ramp prediction or the application of sophisticated monitoring techniques. Finally, in Sect. 4, conclusions are drawn and future research perspectives are sketched.

2   Framework

The WINDML framework aims at minimizing the obstacles for data-driven research in the wind power domain. It allows to simplify numerous steps like loading and preprocessing large scale wind data sets or the effective parameterization of machine learning and data mining approaches. With a framework that bounds specialized mining algorithms to data sets of a particular domain, frequent steps of the data mining process chain can be re-used and simplified. The WINDML framework is released under the open source *BSD 3-Clause license*.

2.1   Overview

An overview of the WINDML framework architecture is shown in Fig. 1. Data sets that are publicly available on the internet can be parsed via an instance of the `DataSource` class. Along with an associated wind turbine or wind park handled via the `Turbine` and `Windpark` classes, such a source can be mapped to feature-label pairs that are amenable to, e.g., supervised learning methods. The corresponding `Mapping` class greatly simplifies the access to various learning schemes provided by, e.g., the SCIKIT-LEARN [9] framework. The analysis and the visualization of the output obtained by corresponding techniques and models is further simplified by the framework via various functions.

The WINDML framework provides a data server, which automatically downloads requested data sets to a local cache when used for the first time. The system only downloads the data for the requested wind turbines and associated time range. The local copies on the user's hard disk are stored in the NUMPY [17] binary file format, allowing an efficient storage and a fast (recurrent) loading of the data. The data set interface allows the encapsulation of different data sources, resulting in a flexible and extendible framework. A complete documentation describing the modules and the overall framework in more detail is openly available. Further, documentations and examples are available, e.g., for the tasks of wind power prediction or visualization, whose outputs are rendered with text and graphics on the WINDML project website.

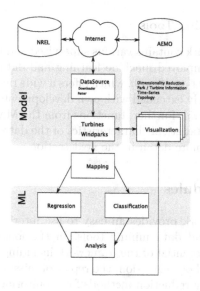

Fig. 1. Overview of WINDML architecture

2.2   Data Sets

WINDML allows the integration of arbitrary wind databases consisting of time series of spatially distributed turbines. The examples presented in this work are based on the *Western Wind Data Set* set [11] of the *National Renewable Energy Laboratory* (NREL), which consists of wind energy time series data of 32,043 wind turbines, each holding ten 3 MW turbines over a timespan of three years in a 10-minute resolution. The data is based on a numerical weather prediction model, whose output has been modified with statistical methods in such a way, that the ramping characteristics are more comparable with those observed in reality [11]. The integration of further data sets is effectively supported by the framework and subject of ongoing work.

2.3   Data Preprocessing

The application of machine learning and data mining tools to raw time series data often requires various preprocessing steps. For example, wind power prediction for a target wind turbine using turbines in the neighborhood affords the composition of the given power values as feature vector matrix. The WINDML framework simplifies such tasks and offers various methods for assembling corresponding patterns. It also effectively supports the imputation of missing data as well as other sophisticated processes including the computation of high-level features. In addition, the framework also provides various small helper functions, which, for instance, address the tasks of calculating the haversine distance between given coordinates or selecting the wind turbines in a specified radius. All these preprocessing functions are provided in the **preprocessing** sub-package.

2.4 Machine Learning Tools

The WINDML framework contains various supervised and unsupervised learning models. Most of the employed machine learning and data mining implementations are based on SCIKIT-LEARN [9], which offers a wide range of algorithms. The methods are continuously extended with own developments such as evolutionary optimization methods [15]. Further, functions from the MATPLOTLIB library are employed to visualize statistical characteristics of the data. Some exemplary use cases are described in more detail in the next section.

3 WindML Modules

The WINDML framework provides an easy-to-use interface for the application of machine learning and data mining tools. In the following, selected examples demonstrate the potential of the framework including basic statistics, wind energy prediction based on regression, the repair of missing data, and the application of dimensionality reduction methods for monitoring tasks.

3.1 Basic Statistics and Visualization

Objective of the basic statistics and visualization modules of WINDML is to support the data mining engineer with information that might be necessary for specific applications. It comprises methods that allow the extraction of simple statistical properties for the development and application of more sophisticated data mining techniques. Various tools and functions are provided by the framework to support this process, including methods for generating features from wind time series, statistical surveys of wind, power, ramps and other properties as well as methods for visualization of time series and topographical characteristics. As an example, consider Fig. 2(a), which shows a histogram plot of wind

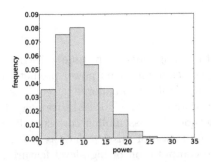

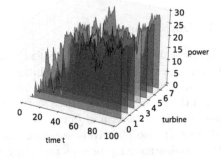

(a) Histogram of wind speeds, *Cheyenne*   (b) Comparison of time series, *Tehachapi*

Fig. 2. (a) Histogram of wind speeds for a turbine in *Cheyenne*, (b) comparison of wind time series of seven turbines near *Tehachapi*

speeds for a turbine near *Cheyenne*. A comparison between the wind time series of seven turbines in a park near *Tehachapi* shows Fig. 2(b).

3.2 Wind Energy Prediction

An important use case is wind energy prediction. The prediction modules are based on spatio-temporal regression models of the wind power time series data for a ten minute to few hour ahead prediction. Our premise is that data-driven models often outperform meteorological models on short prediction horizons. As the strength of meteorological predictions grows with larger time horizons, the hybridization between both is recommendable.

The regression scenario is the following. Given a training set of pattern-label observations $\{(\mathbf{x}_1, y_1), \ldots, (\mathbf{x}_N, y_N)\} \subset \mathbb{R}^d$, we seek for a regression model $f : \mathbb{R}^d \rightarrow \mathbb{R}$ that learns reasonable predictions for power values of a target turbine given unknown patterns $\mathbf{x}'$. We define the patterns as wind power features $\mathbf{x}_i \in \mathbb{R}^d$ of a target turbine and its neighbors at time t (and the past) and the labels as target power values y_i at time $t + \lambda$. Such wind power features $\mathbf{x}_i = (x_{i,1}, \ldots, x_{i,d})^T$ can consist of present and past measurements, i.e., $p_j(t), \ldots, p_j(t - \mu)$ of $\mu \in \mathbb{N}^+$ time steps of turbine $j \in \mathcal{N}$ from the set $\mathcal{N}$ of employed turbines (target and its neighbors) or wind power changes $p_j(t-1) - p_j(t), \ldots, p_j(t - \mu) - p_j(t - \mu + 1)$. Various research questions arise in wind energy prediction. How good is the prediction that we can determine from data and statistical learning models? Strongly connected is the purpose of our prediction, i.e., if it is allowed to over- or underestimate the generated energy. Does the prediction have to avoid large deviations from the power curve or is a good prediction in average acceptable?

The prediction considered in the following regression example concentrates on the prediction of wind power of a single target turbine for a short time horizon λ. The model makes predictions exclusively based on present and past wind power measurements of the target turbine and its neighbors. We employ the present and $\mu = 1$ past measurements and the corresponding difference resulting in $d = (|\mathcal{N}|)(2\mu + 1)$ dimensions. The example considered here is the prediction of power production of a single target turbine in *Palm Springs* (USA, CA, turbine ID 1175) for a time horizon of $\lambda = 3$ (i.e. 30 min). Within a radius of 15 km, the target employs 50 neighbored turbines, from which we select $|\mathcal{N}| = 15$ to construct the input patterns. We train the models on the first half of year 2006, where we employ 3-fold cross-validation to find the best parameters for the regression techniques. The second half of the year is used for evaluation by determining the square error of the forecasts $f(\mathbf{x}_i)$ and the target power value y_i, i.e. $E = \sum_i \left(f(\mathbf{x}_i) - y_i\right)^2$. For comparison of the regression models, we consider $\mu = 2$ time steps, i.e., $d = 30$ features from the time series (two power outputs and their difference for each turbine). The results are compared to the persistence model (PST) $f(t + \lambda) = p(t)$ that assumes that the wind will not change within time horizon λ. Figure 3 shows an experimental comparison between nearest neighbor (kNN) regression and support vector regression (SVR).

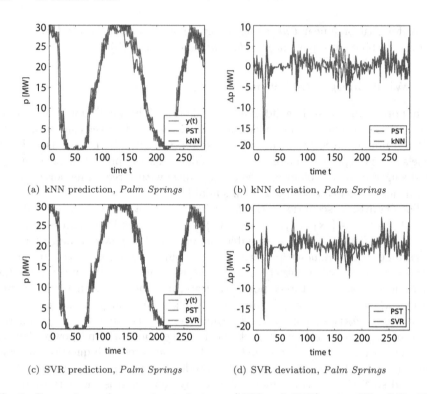

(a) kNN prediction, *Palm Springs*

(b) kNN deviation, *Palm Springs*

(c) SVR prediction, *Palm Springs*

(d) SVR deviation, *Palm Springs*

Fig. 3. Comparison of regression techniques (kNN and SVR) using WINDML. The interval shows an arbitrarily chosen section with a time span of 48 h. The plots show the prediction curves (left) and the deviation from the real power curve (right).

The first regression model we employ for comparison is uniform kNN with Euclidean distance. Since the number of considered neighbors determines the shape of the resulting regression function, we test different values of k, i.e., $k \in \{3, 10, 15, 20, 25, 30, 50, 100\}$ with cross-validation. The best model achieves an error of $E = 174,971.4\,[MW^2]$ with $k^* = 20$, see Table 1. Again, predictions and deviations are shown in Figs. 3(a) and (b). We employ an SVR with RBF kernel and perform grid search for the regularization parameter C and the kernel bandwidth σ testing $C, \sigma \in \{10^{-5}, \ldots, 10^5\}$. The optimal model with $C^* = 10^3$ and $\sigma^* = 10^{-5}$ achieves an error of $E = 140,350.1\,[MW^2]$, see Table 1, and thus achieves the highest accuracy for the tested methods, also compare Figs. 3(c) for the prediction and 3(d) for the deviations. To conclude, both methods outperform the persistence model PST w.r.t. the quadratic error.

In our preliminary work, we proposed the spatio-temporal approach for support vector regression in [7]. We compared regression techniques on the park level in [14]. An ensemble approach has been proposed in [3]. An evolutionary approach for feature (i.e. turbine) selection has been proposed in [15].

Table 1. Comparison of regression methods for wind power prediction.

Method	Optimal parameters	$E(MW^2)$
PST	–	185,453.2
kNN	$k^* = 20$	174,971.4
SVR	$C^* = 10^3$; $\sigma^* = 10^{-5}$	140,350.1

3.3 Ramp Prediction

While the regression model results in a continuous wind power prediction curve, qualitative prediction concentrates on wind increases and decreases. An important use case is the prediction of wind energy ramp events that are large changes of wind energy that play an important role for control strategies in smart grids. In literature, ramps are not clearly defined [6] and may vary from turbine to turbine depending on locations and sizes (for parks respectively). We define power ramp events via a problem-dependent threshold[2] θ. The wind ramp power prediction module treats the ramp prediction problem as classification problem and employs classification methods like support vector machines or decision trees for ramp prediction. A precise ramp classification implies few false positives (missed ramps) and false negatives (false ramps are predicted). This is a difficult challenge, as the problem suffers from imbalanced data due to the comparatively rare ramp events.

In the following ramp prediction example, we treat the ramp prediction problem as three class classification problem and employ a 2-fold cross-validation. Let $p(t)$ be the wind power time-series of the target turbine, for which we determine the forecast. A ramp event is defined as a wind energy change from time step $t \in \mathbb{N}$ to time step $t + \lambda$ with $\lambda \in \mathbb{N}$ by ramp height $\theta \in (0, p_{max}]$, i.e., for a ramp-up event ($y = 1$), it holds $p(t + \lambda) - p(t) > \theta$, for a ramp-down event ($y = -1$) it holds $p(t + \lambda) - p(t) < -\theta$. Non-ramp events get the label $y = 0$.

Figure 4 shows the cross-validation score using recursive feature elimination (RFE) [2] with a linear SVM for the turbines (a) in *Palm Springs* and (b) in *Reno* for two ramp heights and prediction horizons w.r.t. a varying number of features, which have been recursively eliminated from the learning setting. For time critical applications, the reduction of the number of relevant features is an important aspect. The plots show that the cross-validation score is decreasing with increasing number of features from one to ten. In many situations, the smallest cross-validation score is achieved with a feature subset. On *Palm Springs*, the model quality differs significantly. The best results have been achieved for $\theta = 15, \lambda = 2$, where the error is minimal as of about 40 features. Adding further features leads to slight deteriorations, i.e., there exists an optimal subset of features that is remarkably smaller than the maximum number of considered

---

[2] Similarly, Greaves *et al.* [1] define a ramp event for a wind farm as a change of energy output of more than $\theta = 50$ of the installed capacity within a time span of four hours or less.

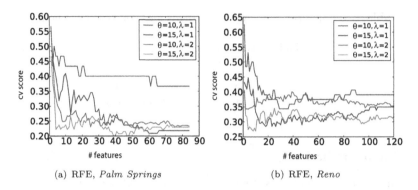

(a) RFE, *Palm Springs*   (b) RFE, *Reno*

Fig. 4. RFE for ramp prediction [8] with a linear SVM with 2-fold cross-validation for (a) *Palm Springs*, and (b) *Reno* for $\theta = 10, 15$ and $\lambda = 1, 2$.

wind turbines. Also on *Reno*, there is a minimum for two of four models ($\lambda = 2$), while the cross-validation error is again increasing and later approximately constant for a larger number of features. In our preliminary work, we analyzed the ramp prediction problem in detail, in particular w.r.t. imbalanced data sets and dimensionality reduction preprocessing [8].

3.4 Data Repair

In data mining applications, patterns are often incomplete, e.g., due to sensor failures. The prediction models often fail in working properly with incomplete or destroyed patterns. Objective of the data repair modules is to employ imputation to fill gaps in patters with missing entries. An inappropriate method might introduce a bias. The research questions related to data repair tasks are: Which method is appropriate for wind energy time series imputation? How can a bias caused by imputation be avoided? Can spatio-temporal relations be exploited for better data repair?

In the following, we demonstrate the data repair approach for wind power time series that are uniformly destroyed, i.e., missing at random (MAR). For this sake, the training data of a target turbine are artificially destroyed for a turbine in *Palm Springs* and a turbine in *Reno*. The training data consist of nine months in 2004. In the first step, the training set is destroyed employing MAR. Then, the repair methods are applied to complete patterns with missing entries. Prediction models are trained with the completed training set and forecasts are computed on a test data sets for nine months in 2005. Figure 5 shows the comparison of regression techniques w.r.t. the prediction accuracy and an increasing rate of incomplete test patterns. The red line shows a linear imputation method, the cyan line the naive model called last observation carried forward (LOCF). The green line shows a multi-linear regression, and the blue line imputation with kNN. The repair with kNN achieves the best imputation results w.r.t. the MSE. For a more detailed comparison of repair techniques, we refer to [10].

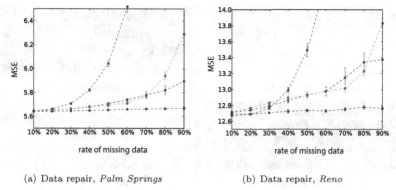

(a) Data repair, *Palm Springs*   (b) Data repair, *Reno*

Fig. 5. Comparison of imputation methods [10], i.e., linear imputation (red), LOCF (red), multi-linear regression (green), and kNN regression (blue) w.r.t an increasing data missing rate for (a) *Palm Springs* and (b) *Reno* (Color figure online).

3.5   Monitoring with Dimensionality Reduction

In WINDML, the monitoring of high-dimensional wind power time series patterns is achieved with dimensionality reduction (DR). DR methods map high-dimensional patterns $\mathbf{X} = \{\mathbf{x}_i \in \mathbb{R}^d\}_{i=1}^N$ to low-dimensional representations $\{\hat{\mathbf{x}}_i \in \mathbb{R}^q\}_{i=1}^N$ in a latent space $\mathbb{R}^q$ with $q < d$. Ideally, such a mapping should preserve most of the properties of the original patterns (e.g., topological characteristics like distances and neighborhoods). In wind power time series[3], such properties can be gradual changes caused by changing weather conditions or even seasonal changes on a slower time scale. The visualization of alert states is a further important part of monitoring systems. Various interesting research questions arise in this context. Which are efficient and effective DR methods for real-time monitoring? Can high-dimensional interactions be visualized by dimensionality reduction techniques? Is dimensionality reduction an appropriate preprocessing technique for prediction tasks in wind energy time series?

The monitoring module of WINDML allows embedding into continuous latent spaces with DR methods like isometric mapping (ISOMAP) [13] and locally linear embedding (LLE) [12]. It allows the visualization of DR results along the time axis. For this sake, the latent positions of the trained manifold are used for colorization of a horizontal bar over time for a training time series. After learning, pattern $\mathbf{x}_t$ of time step t of a test time series is assigned to the color that depends on the latent position $\hat{\mathbf{x}}^*$ of its closest embedded pattern $\mathbf{x}^* = \arg\min_{\mathbf{x}' \in \mathbf{X}} \|\mathbf{x}_t - \mathbf{x}'\|^2$ in the training manifold. For training, $N_1 = 2000$ patterns have been used. We visualize a test set of $N_2 = 800$ patterns at successive time steps in the following figures. Figure 6 shows the monitoring results of ISOMAP and LLE with $k = 10, 30$. Areas with a similar color and few color changes can be found in each case. Areas with frequent changes can be observed at the same points of time in each plot. ISOMAP and LEE turn out to be robust

---

[3] We define a pattern as $\mathbf{x}_i = (p_1(t), \ldots, p_d(t))^T$ with turbines $1, \ldots, d$ at time $t = i$.

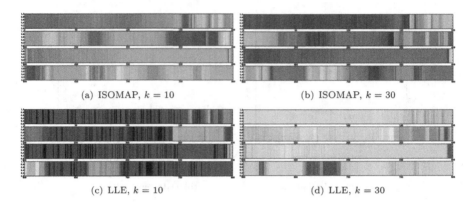

(a) ISOMAP, $k = 10$ (b) ISOMAP, $k = 30$

(c) LLE, $k = 10$ (d) LLE, $k = 30$

Fig. 6. Visualization of a test sequence employing ISOMAP in (a) and (b) LLE in (c) and (d) with varying k for $d = 66$ turbines in *Tehachapi* (Color figure online).

w.r.t. the choice of neighborhood size k. However, the learning result of LLE with $k = 10$ is worse in stable situations showing fluctuating colors (blue and black, see Fig. 6(c)). ISOMAP generates more robust results independent of neighborhood sizes. In [7], we employ self-organizing maps (SOMs) for sequence visualization of high-dimensional wind time series.

4   Conclusion

The WINDML framework is an easy-to-use PYTHON-based software library that allows an effective development of machine learning tools for wind energy time series data. The use cases sketched in this work have an important part to play for the integration of wind energy in today's and future smart grids. The extension of modules, the integration of further wind data sets, and the framework's adaptation to other research and business applications in wind energy information systems are future steps of WINDML. For instance, we plan to extend WINDML by turbine placement strategies that take into account geo-information and wake effects. Further, the development of wind energy prediction techniques based on neural networks and evolutionary computation are part of ongoing work.

Acknowledgments. We thank the *National Renewable Energy Laboratory* for providing the *Western Wind Data Set* [11].

References

1. Greaves, B., Collins, J., Parkes, J., Tindal, A.: Temporal forecast uncertainty for ramp events. Wind Eng. **33**(4), 309–319 (2009)
2. Guyon, I., Weston, J., Barnhill, S., Vapnik, V.: Gene selection for cancer classification using support vector. Mach. Learn. **46**(1–3), 389–442 (2002)

3. Heinermann, J., Kramer, O.: Precise wind power prediction with SVM ensemble regression. In: Wermter, S., Weber, C., Duch, W., Honkela, T., Koprinkova-Hristova, P., Magg, S., Palm, G., Villa, A.E.P. (eds.) ICANN 2014. LNCS, vol. 8681, pp. 797–804. Springer, Heidelberg (2014)
4. Hunter, J.D.: Matplotlib: a 2d graphics environment. Comput. Sci. Eng. 9(3), 90–95 (2007)
5. Jones, E., Oliphant, T., Peterson, P., et al.: SciPy: open source scientific tools for Python (2001). Accessed 15 July 2014
6. Kamath, C.: Understanding wind ramp events through analysis of historical data. In: Proceedings of the IEEE PES Transmission and Distribution Conference and Exposition, pp. 1–6 (2010)
7. Kramer, O., Gieseke, F., Satzger, B.: Wind energy prediction and monitoring with neural computation. Neurocomputing 109, 84–93 (2013)
8. Kramer, O., Treiber, N.A., Sonnenschein, M.: Wind power ramp event prediction with support vector machines. In: Polycarpou, M., de Carvalho, A.C.P.L.F., Pan, J.-S., Woźniak, M., Quintian, H., Corchado, E. (eds.) HAIS 2014. LNCS (LNAI), vol. 8480, pp. 37–48. Springer, Heidelberg (2014)
9. Pedregosa, F., Varoquaux, G., Gramfort, A., Michel, V., Thirion, B., Grisel, O., Blondel, M., Prettenhofer, P., Weiss, R., Dubourg, V., Vanderplas, J., Passos, A., Cournapeau, D., Brucher, M., Perrot, M., Duchesnay, E.: Scikit-learn: machine learning in Python. J. Mach. Learn. Res. 12, 2825–2830 (2011)
10. Poloczek, J., Treiber, N.A., Kramer, O.: KNN regression as geo-imputation method for spatio-temporal wind data. In: de la Puerta, J.G., et al. (eds.) International Joint Conference SOCO'14-CISIS'14-ICEUTE'14. AISC, vol. 299, pp. 185–193. Springer, Heidelberg (2014)
11. Potter, C.W., Lew, D., McCaa, J., Cheng, S., Eichelberger, S., Grimit, E.: Creating the dataset for the western wind and solar integration study (USA). In: 7th International Workshop on Large Scale Integration of Wind Power and on Transmission Networks for Offshore Wind Farms, (2008)
12. Roweis, S.T., Saul, L.K.: Nonlinear dimensionality reduction by locally linear embedding. Science 290, 2323–2326 (2000)
13. Tenenbaum, J.B., Silva, V.D., Langford, J.C.: A global geometric framework for nonlinear dimensionality reduction. Science 290, 2319–2323 (2000)
14. Treiber, N.A., Heinermann, J., Kramer, O.: Aggregation of features for wind energy prediction with support vector regression and nearest neighbors. In: European Conference on Machine Learning (ECML), Workshop DARE (2013)
15. Treiber, N.A., Kramer, O.: Evolutionary turbine selection for wind power predictions. In: Lutz, C., Thielscher, M. (eds.) KI 2014. LNCS, vol. 8736, pp. 267–272. Springer, Heidelberg (2014)
16. Vanderplas, J., Connolly, A., Ivezić, Ž, Gray, A.: Introduction to astroml: machine learning for astrophysics. In: Conference on Intelligent Data Understanding (CIDU), pp. 47–54 (2012)
17. van der Walt, S., Colbert, S.C., Varoquaux, G.: The numpy array: a structure for efficient numerical computation. Comput. Sci. Eng. 13(2), 22–30 (2011)

Systematical Evaluation of Solar Energy Supply Forecasts

Robert Ulbricht[2]($\boxtimes$), Martin Hahmann[1], Hilko Donker[2], and Wolfgang Lehner[1]

[1] Database Technology Group, Technische Universität Dresden, Dresden, Germany
[2] Robotron Datenbank-Software GmbH, Dresden, Germany
robert.ulbricht@robotron.de

Abstract. The capacity of renewable energy sources constantly increases world-wide and challenges the maintenance of the electric balance between power demand and supply. To allow for a better integration of solar energy supply into the power grids, a lot of research was dedicated to the development of precise forecasting approaches. However, there is still no straightforward and easy-to-use recommendation for a standardized forecasting strategy. In this paper, a classification of solar forecasting solutions proposed in the literature is provided for both weather- and energy forecast models. Subsequently, we describe our idea of a standardized forecasting process and the typical parameters possibly influencing the selection of a specific model. We discuss model combination as an optimization option and evaluate this approach comparing different statistical algorithms against flexible hybrid models in a case study.

Keywords: Solar energy · Energy forecast model · Classification · Ensemble

1 Introduction

The capacity of renewable energy sources (RES) constantly increases world-wide due to governmental funding policies and technological advancements. Unfortunately, most of the grid-connected RES installations are characterized by a decentralized allocation and a fluctuating output owed to the changing nature of the underlying powers. Coincidentally, today's available transformation and storage capabilities for electric energy are limited and cost-intensive, which is the primary reason for the increasing interference of renewable energy output with power network stability. Efficient and dedicated forecasting methods will help the grid operators to better manage the electric balance between power demand and supply in order to avoid unstable situations or even possible collapses in the near future. A lot of research has been conducted in the past years by different communities trying to cope with this challenge. Despite of the large amount of available related work and both scientific and practical optimization ideas, there is still no straightforward and easy-to-use recommendation for a standardized forecasting strategy. Comparing the results obtained while executing different

© Springer International Publishing Switzerland 2014
W.L. Woon et al. (Eds.): DARE 2014, LNAI 8817, pp. 108–121, 2014.
DOI: 10.1007/978-3-319-13290-7_9

experimental approaches is difficult, as most of the presented cases are bound to a specific region including the corresponding real-world data-set. Further, there is no constant form of result evaluation across all publications, as different error metrics are applied to measure output quality.

In this paper, we address the problem of a systematical optimization for solar energy forecasting strategies conducting an analysis of state-of-the-art approaches. The paper is organized as follows: In Sect. 2 we review and classify models proposed in the literature to predict (1) weather influences and (2) the output of solar energy production units. In Sect. 3, the energy forecasting process is described and relevant parameter settings and exogenous influences for the model selection decision are discussed before that background. In Sect. 4 we evaluate the performance of an exemplary ensemble model which combines the forecast output of popular statistical prediction methods using a dynamic weighting factor. Finally, we conclude and outline additional research directions for our future work in Sect. 5.

2   Energy Supply Forecasting Approaches

The prediction of energy time series is a classical application of time series analysis methods. Thus, there is a long research history related to electricity load forecasting, where a range of sophisticated high-quality models has been developed and classified (i.e. compare the work of Alfares and Nazeeruddin [2]). In contrast, the need for energy supply forecasting is a much more recent topic, as the challenge of grid-connected RES penetrating the distribution systems has emerged just a couple of years ago. Nevertheless, both energy demand and supply forecasting approaches make use of similar techniques.

2.1   Weather Forecast Models

In order to make energy supply planning rational, forecasts of RES production have to be made based on the consideration of weather conditions as the most influencing factor for output determination for solar energy production is the quality of the solar irradiation forecast. Consequently, the use of precise weather forecast models is essential before reliable energy output models can be generated. Although this step is orthogonal to a grid operator's core activities (weather data usually is obtained from meteological services), a basic understanding of the underlying principles is helpful when choosing a specific energy output model.

Numerical Weather Prediction. Complex global *numerical weather prediction* (NWP) models is a modern and common method to predict a number of variables describing the physics and dynamic of the atmosphere, which is then used to derive the relevant weather influences at a specific point of interest. These are e.g. the *European Center for Medium-Range Weather-Forecasts Model*[1] (ECMWF), the *Global Forecast System* (GFS) from National Centers for

---

[1] http://www.ecmwf.int

Environmental Prediction[2] or the *North American Mesoscale Model*[3] (NAM). As they have a coarse spatial and temporal resolution, several post-processing and correction techniques are applied in order to obtain down-scaled models of finer granularity (e.g. Model Output Statistics). A quality benchmark was conducted by Lorenz et al. [15], where European ground measurement data is used to compare the performance of each NWP including different scientific and commercial post-processing approaches (Fig. 1).

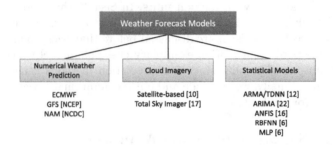

Fig. 1. Classification of weather forecasting models

Cloud Imagery. The influence of local cloudiness is considered to be the most critical factor for the estimation of solar irradiation, especially on days with partial cloudiness where abrupt changes may occur. The use of satellite data can provide high quality short-term forecasts, as geostationary satellites like METEOSAT provide half-hourly spectrum images with a resolution from 1 to 3 square kilometers. Clouds are detected by processing these images into cloud-index images. To predict the future position of a cloud over ground, two consecutive cloud-index images are interpolated using motion vectors [10]. A similar method is the use of *Total Sky Imagers*, which enables real-time detection of clouds in hemispherical sky images recorded by ground-based cameras using sophisticated analytical algorithms [17].

Statistical Models. Furthermore, there are several studies treating the forecasting of solar radiation based on historical observation data using common time series regression models like ARIMA, *Artificial Neural Networks* (ANN) or *Fuzzy-Logic* models (FL). An analysis published by Reikard shows that after comparing various regression models, ARIMA in logs with time-varying coefficients performs best, due to its ability to capture the diurnal cycle more effectively than other methods [22]. Ji and Chee [12] propose a combination of ARMA and a *Time Delay Neural Network* (TDNN). Dorvlo et al. discuss the usage of two ANN-models: *Radial Basis Functions* (RBF) and *Multilayer*

---

[2] http://www.ncep.noaa.gov
[3] http://www.ncdc.noaa.gov

Perceptron (MLP) [6]. Martin et al. [16] compare the performance of auto-regressive models (AR) against ANN and Adaptive-network-based fuzzy inference system (ANFIS). As such statistical models usually are considered being domain-neutral, their characteristics are discussed more in detail in the subsequent section.

2.2 Energy Forecast Models

Any output from the weather models described above must then be converted into electric energy output. According to the underlying methodology, the existing solutions can be classified into the categories of *physical*, *statistical* and *hybrid* methods as presented in Fig. 2.

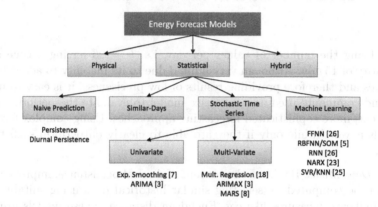

Fig. 2. Classification of energy forecasting models

Physical Models. All forecasting approaches mainly relying on a renewable power plant's technical description concerning its ability to convert the introduced meteorological resources into electrical power are summarized by the term *physical model*. Taking into account external influences derived from NWP, atmospheric conditions and local topography, once they are fitted they are accurate and do not require historical output curves. Especially the latter makes them suitable for estimating the future output of planned or recently installed RES units. Applications of physical models are more frequently found for wind power prediction, but are also used for solar energy forecasts. For example, if we consider the electrical energy P_E extracted from the NWP for global radiation G_{nwp} by a PV panel, the equation for a simplyfied model is as follows:

$$P_E = \alpha G_{nwp} A \tag{1}$$

where α is the conversion efficiency of the solar panel and A is its surface size. Improvements of this method are demonstrated by Iga and Ishihara [11] including the outside air temperature, or Alamsyah et al., using the panel temperature

[1] as additional parameters. The major disadvantage of physical models is that they are highly sensitive to the NWP prediction error. Furthermore, they have to be designed specifically for a particular energy system and location. As a consequence, the usage of such models requires detailed technical knowledge about characteristics and parameters of all underlying components, thus making them more relevant for energy plant owners or producers than for grid operators.

Statistical Models. *Naive Prediction.* The most straightforward approach to determine a time series' future value denoted as P'_{t+1} would be a naive guess, assuming that next periods' expected energy output will be equal to the observations of the current period P_t. This method is called *naive* or *persistent prediction.* The distinctive cycle of solar energy is expressed by choosing a period of 24 h for *diurnal persistence*, so forecasts are obtained by

$$P'_t = P_{t-k} \tag{2}$$

with k being the number of values per day, i.e. $k = 96$ having a time series granularity of 15 min. Although very limited due to its inability to adopt to any influences and therefore providing results of low preciseness, it is easy to implement and commonly used as a reference model to evaluate the performance of concurrent, more sophisticated forecasting approaches. Using complex forecasting tools is worthwhile only if they are able to clearly outperform such trivial models.

Similar-Days Model. Based on the concept of diurnal persistence, improved forecasts can be computed by selecting similar historical days using suitable time series similarity measures like e.g. Euclidean distance. These models are very popular for load forecasts (e.g. compare [19]), where weather-awareness plays a minor part compared to the influence of consumption-cycle patterns derived from historical data. As for solar energy forecasts, such models are used whenever there is no NWP available at all or the prediction error included naturally in the NWP is estimated as too high to provide reliable energy output forecasts.

Stochastic Time Series. Depending on the number of influencing parameters, two groups of models can be distinguished: *Uni-* and *Multivariate* models. Univariate models are calculated based on the time series' history only. Well known representatives of that group are *Auto-Regressive (Integrated) Moving Average* models (ARMA/ARIMA), which can be described best as a stochastic process combining an auto-regressive component (AR) with a moving average component (MA). Dunea et al. [7] propose the consideration of *Exponential Smoothing* as an effective alternative for one-period ahead forecasts. In contrast, multivariate models allow for the integration of exogenous parameters. *Multiple Regression* methods like ARIMAX (ARIMA with exogenous influences) are a popular choice whenever there is a linear correlation structure expected in two time series [18]. In the case of solar energy prediction, this is given by the dominating dependency of energy output on the global radiation values from the NWP. Historical observation data is used to derive the regression coefficients. Bacher et al. demonstrate the performance of an ARIMA model using a clear-sky-normalization for short-term forecasts [3]. As an extension to linear modeling *Multivariate Adaptive Regression Splines* (MARS), a methodology developed by Friedman [8], is used in the energy domain to generate more flexible, nonlinear models.

Machine Learning. The use of machine learning methods is a common approach to forecast a time series' future values, as they are seen as alternative to conventional linear forecasting methods. Reviewed literature shows that ANN have been successfully applied for forecasts of fluctuating energy supply. ANN learn to recognize patterns in data using training data sets. For example, the use of neural networks is proposed by Yona et al. [26] due to their examination of the *Feed-Forward* (FFNN), the *Radial Basis Function* (RBFNN) and the *Recurrent Neural Network* (RNN) for solar power forecasting based on NWP input and historical observation data. A similar approach is described by Chen et al. [5], where a RBFNN is combined with a weather type classification model obtained by a *Self Organizing Map* (SOM). Wolf et al. compare *k-Nearest Neighbors* (KNN) and *Support Vector Regression* (SVR) finding that the latter outperforms KNN on non-aggregated data [25]. In contrast, Tao et al. compute hourly energy forecasts using an adaptive NARX network combined with a clear-sky radiation model, which allows for forecasts without including NWP data and still outperforms non-adapting regression-based methods [23].

Hybrid Models. Any combination of two or more of the above described methods is known as a *hybrid model*. The use of such hybrid approaches has become more popular as it offers the possibility to take advantage of the strongest points of different stand-alone forecasting techniques. The basic idea of combining models is to use each methods' unique features to capture different patterns in the data. Theoretical and empirical findings from other domains suggest that combining linear and non-linear models can be an efficient way to improve the forecast accuracy (e.g. [27]), so hybrid models seem to be a promising approach that can potentially outperform non-hybrid models individually. A successful application of this idea is provided e.g. by the work of Ogliari et al. [21].

3   Energy Forecasting Process

As shown in the previous section, there are plenty of possibilities to compute forecasts for fluctuating energy production units. But choosing the optimal forecasting model for a given use case is an important decision to make and requires expert knowledge. Figure 3 describes the forecasting process steps: First, the raw data has to be *preprocessed* according to the specific requirements of the wanted forecast. Second is the *selection* of a suitable algorithm to best describe the observations. Next, the parameters for the chosen model have to be *estimated* before the *forecasting* task is executed. After *evaluating* the obtained results, this decision might be reconsidered in case of too high and therefore unsatisfying prediction errors. From the description of the forecasting techniques mentioned in

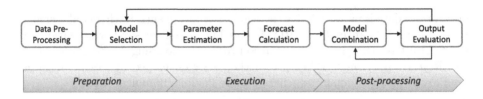

Fig. 3. Typical forecasting process steps

the introduction, we derive that the choice of the appropriate forecasting models depends on the amount and quality of external information, the applied forecast horizon, the data aggregation level and the availability of historical observation data. Furthermore, we consider *model combination* as an optimization option, a strategy also known as *ensemble prediction*. Ensembles can be created manually based on user preferences and experiences or by using machine driven optimization approaches. However, another fact to consider when choosing among forecasting models is their efficiency: There is an economic preference for inexpensive and easy-to-use methods if they promise satisfying results.

Context Information. The availability of weather forecasts is an essential condition for both physical and multiple-regression forecasting models, most importantly the quality of solar irradiation values. Predicted outside air temperature can be used to estimate a panel's future surface temperature, as energy output is reduced significantly on hot cells. In a similar manner, wind speed can indicate cooling effects. Further, technical information like the panels inclination angle and production year (due to the fact that their conversation efficiency decreases over age) are interesting. As for environmental influences, cleaning cycles are relevant because polluted panels will produce significantly less energy, which is a considerable influence in dry and dusty areas. Also, in some regions, detected snow coverage might prevent any energy output at all.

Forecast Horizon. Studies show that the forecast horizon for which a generated model has to be applied is an important parameter while choosing an appropriate prediction approach. In the energy domain, forecast horizons are determined depending on the requirements of the underlying business process. Usually, we can distinguish the following categories: now-casts (up to 4 h ahead), short-term (up to 7 days ahead) and long-term forecasts (more than 7 days ahead). Focusing on the grid operators activities related to renewable energy integration, we find that intra-day and day-ahead horizons represent the most relevant time scales for operations [22], while long-term predictions are of special interest for resource and investment plannings.

Spatial and Temporal Aggregation. Forecast quality of statistical or physical models will vary strongly depending on the level of spatial aggregation of the underlying energy time series. Since it is well known that the overall impact of single peak values decreases with increasing size of the chosen aggregates, forecasts computed on single or disaggregated time series usually contain the risk of higher average prediction errors. In contrast, following an agglomerative or bottom-up approach by creating different aggregation levels might lead to better results on higher levels (e.g. average cloudiness in a region can be predicted more accurately than cloudiness at a particular site [14]), but complicates the integration of available context information, especially in the case of influences characterized by strong locality. The use of clustering techniques to create hierarchical aggregations of RES time series is a matter of a separate study [24] in progress. Temporal aggregation can be considered if the granularity of source time series needs to be reduced in order to obtain forecasts of lower resolution.

History Length. Stochastic approaches create a forecast model over the historical supply curves. The size of available history influences the accuracy of the forecasting result, as a longer history length might be suitable for learning repeatable patterns, while a shorter history length is more beneficial for strongly fluctuating time series. The latter requires a continuous adaption of the forecast models and, possibly, also of the history length. However, determining the best model parameters involves multiple iterations over the time series history which is an expensive process especially on large data sets. Reducing the history length can therefore speed up model creation significantly. Previous research in this area [9] proposes an I/O-conscious skip list data structure for very large time series in order to determine the best history length and number of data points for linear regression models.

4 Model Selection - A Case Study

In this section we analyze the impact of the previously described energy model selection parameters on the forecast output. After briefly introducing the forecasting methods to be assessed, we provide a description of our experimental setting including the used data set, the applied methodologies and the output evaluation criteria before we discuss the obtained results.

4.1 Predictor Description

Several forecasting algorithms have been chosen for our evaluation: (1) The Similar-Days model using Euclidean distance and (2) the univariate *Autoregressive Fractionally Integrated Moving Average* (ARFIMA) model that both are weather-unaware. Regression-based models are represented by (3) Mirabel[4], a scientific model based on principal component analysis and multivariate regression and (4) Epredict[5], a commercial library using the non-linear MARS algorithm. Additionally, (5) a domain-neutral multiple linear regression model from the OpenForecast[6] library is included. Hence, all classes of statistical models are covered except machine learning. The benchmark will be conducted against a naive model using diurnal persistence.

---

[4] http://www.mirabel-project.eu/

[5] http://www.robotron.eu/

[6] http://www.stevengould.org/software/openforecast/

Table 1. Sample data properties

Time series	Aggregation level (extension)	Peak P_{max}	Installations (capacity)
DIA	None	42.75 kW	1 (351.3 kW)
DSA	Distribution system (23 km^2)	257.45 kW	7 (3,440 kW)
TSA	Transmission system (109,000 km^2)	1,616.04 MW	107,216 (7,008 MW)

4.2 Methodology

The Data. To cope with the recently introduced model selection criteria of spatial aggregation, we include three observed solar energy output curves into our scenario: (1) A single, disaggregated PV-installation located in central Germany denoted as *DIA*, (2) an aggregate built of all measured PV-installations available in the same local distribution system denoted as *DSA* and (3) an aggregate build of all PV-installations attached to the superior transmission system denoted as *TSA*. DIA and DSA were provided by a cooperating distribution system operator[7], while TSA was obtained from a public website[8]. All time series have a resolution of 15 min and cover all of the year 2012. Corresponding weather data including measurements of solar irradiation, air temperature, and wind speed with a resolution of 1 h is available from a weather station run by a meteorological service[9], located within the distribution networks' range. Using weather observations instead of weather forecasts eliminates the naturally included NWP prediction error thus allowing for a unbiased evaluation of the energy model performance itself (Table 1).

Operational Framework. We use the first 11 months of historical data from our source time series for model training. Forecasts are computed for the remaining month, thus providing a test data set of 2976 predicted values according to our time series' resolution. To cover both intra-day and day-ahead terms with our scenario, we define varying forecast horizons of 2, 12 and 24 h ahead. After computing a forecast the training period is adopted by adding the forecast horizon length, thus extending the available history length accordingly with each completed iteration. A suchlike *moving origin* approach simulates the integration of newly arriving observations in the forecasting process, which can then be compared with the latest forecast output and used to adjust the forecast model. Therefore, the number of forecasting models required to cover the whole month is 372 for a horizon of 2 h, 62 (12 h) and 31 (24 h) respectively. Finally, the test data is split into a calibration, and an evaluation period.

Combination Approach. Various forms of combining forecasts have been developed like subjective human judgment or objective approaches, the latter

---

[7] http://www.en-apolda.de
[8] http://www.50hertz.com
[9] http://wetterstationen.meteomedia.de

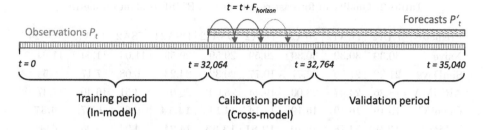

Fig. 4. Operational benchmark framework using moving origin

extending from the simple arithmetic mean to more sophisticated methods such as neural networks. A classification and identification of the most common methods can be found in [4]. In our study, we apply an objective method to combine the forecasts through a linear combination of n non-biased individual forecasts using an unrestricted weight factor λ_n for each forecast. The final energy forecast P_t' is then computed by

$$P_t' = \sum_1^n \lambda_n P_{nt}' \tag{3}$$

where P_{nt}' is the forecasted value from model n for a timestamp t. In order to derive the optimal weight factors we use Nelder-Mead function minimization [20], which aims at reducing the error in the forecast output during the calibration period. After experimenting with different sets of input parameters, best results were obtained using the RMSE as target function and a calibration period of 700 values (approx. 1 week) as depicted in Fig. 4. Finally, several ensembles were computed using the n-best models in terms of RMSE denoted as *Ensble-nB* or all available individual models denoted as *Ensble-All*.

Output Evaluation. To evaluate the quality of the predicted values, different statistical accuracy metrics can be used for illustrating either the systematic or random errors included in the results. The *root mean square error* (RMSE) is the recommended measure and main evaluation criterion for intra-day forecasts, as is addresses the likelihood of extreme values better [13]. As the RMSE returns absolute values, normalization is applied in order to allow for model performance comparison on time series having different aggregation scales. The *normalized root mean square error* (nRMSE) is achieved by

$$nRMSE = \frac{100}{P_{max}} * \sqrt{\frac{\sum_{t=1}^n (P_t - P_t')^2}{n}} \tag{4}$$

with P_{max} being the maximum observed power output in the validation data set. Beside the nRMSE, the systematic error can be expressed by computing the average error over the whole evaluation period. The *normalized mean bias error* (nMBE) is found by

$$nMBE = \frac{100}{P_{max}} * \frac{\sum_{t=1}^n (P_t - P_t')}{n} \tag{5}$$

Table 2. Quality of forecast results using nRMSE evaluation metric

Predictor	DIA2	DIA12	DIA24	DSA2	DSA12	DSA24	TSA2	TSA12	TSA24
Naive	30.39	30.39	30.39	29.59	29.59	29.59	11.04	11.04	11.04
SimDays	21.28	21.59	26.77	19.77	20.32	24.23	6.08	7.17	8.54
ARFIMA	18.26	24.47	23.00	19.44	28.09	26.01	3.51	16.40	17.37
OpenFC	16.19	16.19	16.19	14.14	14.14	**14.14**	6.67	6.67	**6.67**
MARS	**13.99**	14.78	16.10	**12.81**	**13.33**	14.24	4.94	5.96	6.96
Mirabel	15.96	15.96	15.96	14.58	15.57	17.85	3.73	6.55	7.86
Ensble-2B	14.33	**14.61**	**15.71**	15.23	16.68	18.22	**2.49**	5.56	7.57
Ensble-3B	14.14	15.47	15.80	14.82	16.75	18.61	2.54	5.50	7.27
Ensble-4B	14.29	15.66	18.86	14.45	16.73	20.25	2.61	5.15	7.13
Ensble-All	14.25	16.18	18.12	14.56	17.03	18.12	2.62	**5.12**	7.05

and can be used to detect a systematic bias in the forecast, as according to Eq. 5 negative values represent over-estimations and vice versa. Note that non-daylight hours (values with timestamps before 8 am and after 4 pm) and all resting zero observation values are excluded from error calculation. The latter also implies that the effects of snow coverage or measurement failures are removed completely from the results.

4.3   Experimental Results

Our results listed in Table 2 show that in terms of RMSE, almost all forecasting models clearly outperform the naive benchmark with two exceptions being ARFIMA on TSA12 and TSA24. It is also visible that the uni-variate stochastic models SimDays and ARFIMA perform rarely better than those able to integrate external influences, especially on time series with lower aggregation levels. Regarding the impact of the chosen forecast horizon, we observe that the fitness of most of the models is decreasing with longer horizons (compare Fig. 5). In contrast, OpenFC seems to be completely unaffected by that parameter and provides constant values, thus leading to good results for day-ahead forecasts. We suspect that OpenFC can even outperform the sophisticated energy predictors MARS and Mirabel on short- and mid-term forecasts, which have not been covered by the presented scenario.

An analysis of the combined models shows that both improvements and degradations of individual results were obtained. The best results were provided using the 2 best models denoted as Ensble-2B: The RMSE of the best individual model could be reduced by 28.95 % for the TSA2 forecast (compare Fig. 6) and slight improvements were obtained on DIA12 and DIA24 with 1.11 % and 1.60 % respectively. Ensble-3B and Ensble-4B only outperformed the individual models once with the TSA12 forecast. According to expectations all evaluated models show the lowest preciseness on the individual PV-installation denoted as DIA,

Table 3. Quality of forecast results using nMBE evaluation metric

Predictor	DIA2	DIA12	DIA24	DSA2	DSA12	DSA24	TSA2	TSA12	TSA24
Naive	4.45	4.45	4.45	4.33	4.33	4.33	0.69	0.69	0.69
SimDays	0.21	−2.55	−9.02	−0.37	−2.03	−6.01	0.07	−0.11	−0.91
ARFIMA	1.24	−0.97	7.93	**0.03**	−1.96	12.24	−0.85	−11.59	−11.44
OpenFC	4.02	4.02	4.02	0.30	**0.30**	**0.30**	−2.57	−2.57	−2.57
MARS	1.87	2.34	3.97	−0.36	−0.98	−0.47	−1.18	−2.32	−2.74
Mirabel	3.80	3.80	3.80	−3.48	−3.79	−4.17	**−0.06**	−0.60	**−0.35**
Ensble-2B	**−0.06**	2.39	**3.51**	1.08	1.72	2.02	−0.27	−1.39	−1.98
Ensble-3B	0.07	2.72	3.69	0.24	1.68	2.41	−0.27	−0.93	−1.52
Ensble-4B	0.20	2.23	4.27	0.24	1.61	5.37	−0.36	−0.45	−1.10
Ensble-All	−1.09	**0.49**	3.69	0.33	1.00	5.09	−0.38	**−0.05**	−0.65

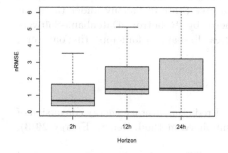

Fig. 5. Distribution of nRMSE values for varying horizons on TSA time series using MARS predictor

Fig. 6. Performance of Ensemble-2B and the underlying individual forecasts on TSA time series and 2h-ahead horizon

since single outliers can hardly be compensated as in the case of aggregated data like DSA or TSA. It is also noted that all models perform best on the transmission system level TSA. Although the correlation of weather information observed at only one specific location is not considered being a representative influence on that supra-regional level, the effect of weather-awareness seems to be completely neutralized by the impact of high aggregation. However, these results are not reflected using the nMBE evaluation criteria listed in Table 3, where in all cases except DIA24 at least one model could provide nearly unbiased forecasts having a nMBE value close to zero.

5   Conclusions

In this work we have shown that the forecasting of solar energy output is a two-step approach, typically requiring a weather- and an energy-forecast model. As for the energy-forecast, possible choices can be selected amongst physical, statistical, and hybrid models. The selection of an appropriate model depends on the characteristics of the underlying data and the relevant evaluation criteria. Combining models offers additional optimization options whenever there is no model to be found that individually outperforms in all given situations, as demonstrated against the parameters of forecast horizon and spatial aggregation. However, deriving generalizable recommendations regarding the selection of appropriate models or model combinations based only on the evaluated use case remains challenging. For our future work, we think that conducting a more complex and global benchmark covering more state-of-the-art forecasting approaches and additional scenarios will provide useful information on how to systematically select an optimal energy model and might unlock the potential towards establishing industry standards regarding the application of forecasting strategies and output evaluation criteria.

Acknowledgment. The work presented in this paper was funded by the European Regional Development Fund (EFRE) and the Free State of Saxony under the grant agreement number 100081313 and co-financed by Robotron Datenbank-Software GmbH. We thank Claudio Hartmann and Andreas Essbaumer for supporting our work.

References

1. Alamsyah, T.M.I., Sopian, K., Shahrir, A.: Predicting average energy conversion of photovoltaic system in Malaysia using a simplified method. Renew. Energy **29**(3), 403–411 (2004)
2. Alfares, H.K., Nazeeruddin, M.: Electric load forecasting: literature survey and classification of methods. Int. J. Syst. Sci. **33**(1), 23–34 (2002)
3. Bacher, P., Madsen, H., Nielsen, H.A.: Online short-term solar power forecasting. Sol. Energy **83**(10), 1772–1783 (2009)
4. Mancuso, A.C.B., Werner, L.: Review of combining forecasts approaches. Indep. J. Manag. Prod. **4**(1), 248–277 (2013)
5. Chen, C., Duan, S., Cai, T., Liu, B.: Online 24-h solar power forecasting based on weather type classification using artificial neural network. Sol. Energy **85**(11), 2856–2870 (2011)
6. Dorvlo, A.S.S., Jervase, J.A., Al-Lawati, A.: Solar radiation estimation using artificial neural networks. Appl. Energy **71**(4), 307–319 (2002)
7. Dunea, A.N., Dunea, D.N., Moise, V.I., Olariu, M.F.: Forecasting methods used for performance's simulation and optimization of photovoltaic grids. In: IEEE Porto Power Tech Proceedings, Porto, Portugal (2001)
8. Friedman, J.H.: Multivariate adaptive regression splines. Ann. Stat. **19**(1), 1–141 (1991)
9. Ge, T., Zdonik, S., Tingjian, G., Stanley, B.Z.: A skip-list approach for efficiently processing forecasting queries. In: Proceedings of the VLDB Endowment, Auckland, New Zealand, pp. 984–995 (2008)
10. Heinemann, D., Lorenz, E., Girodo, M.: Forecasting of solar radiation. In: Wald, L., Suri, M., Dunlop, E.D. (eds.) Solar Energy Resource Management for Elitricity Generation from Local Level to Global Scale, pp. 83–94. Nova Science Publishers, New York (2006)

11. Iga, A., Ishihara, Y.: Characteristics and embodiment of the practical use method of monthly temperature coefficient of the photovoltaic generation system. IEEJ Trans. Power Energy **126**, 767–775 (2006)
12. Ji, W., Chee, K.C.: Prediction of hourly solar radiation using a novel hybrid model of ARMA and TDNN. Sol. Energy **85**(5), 808–817 (2011)
13. Kostylev, V., Pavlovski, A.: Solar power forecasting performance - towards industry standards. In: 1st International Workshop on the Integration of Solar Power into Power Systems, Aarhus, Denmark (2011)
14. Lorenz, E., Hurka, J., Heinemann, D., Beyer, H.G.: Irradiance forecasting for the power prediction of grid-connected photovoltaic systems. Sel. Top. Appl. Earth Observ. Remote Sens. **2**(1), 2–10 (2009)
15. Lorenz, E., Remund, J., Müller, S.C., Traunmüller, W., Steinmaurer, G., Pozo, D., Ruiz-Arias, J.A., Fanego, V.L., Ramirez, L., Romeo, M.G., et al.: Benchmarking of different approaches to forecast solar irradiance. In: 24th European Photovoltaic Solar Energy Conference, Hamburg, Germany, pp. 1–10 (2009)
16. Martin, L., Zarzalejo, L.F., Polo, J., Navarro, A., Marchante, R., Cony, M.: Prediction of global solar irradiance based on time series analysis: application to solar thermal power plants energy production planning. Sol. Energy **84**(10), 1772–1781 (2010)
17. Martins, I., Ferreira, P.M., Ruano, A.E.: Estimation and prediction of cloudiness from ground-based all-sky hemispherical digital images. In: Proceedings of the 2nd Ambient Computing Colloquium in Engineering and Education, Faro, Portugal (2011)
18. Meyer, E.L., van Dyk, E.E.: Development of energy model based on total daily irradiation and maximum ambient temperature. Renew. Energy **21**(1), 37–47 (2000)
19. Mu, Q., Wu, Y., Pan, X., Huang, L., Li, X.: Short-term load forecasting using improved similar days method. In: Asia-Pacific Power and Energy Engineering Conference (APPEEC), ChengDu, China, pp. 1–4 (2010)
20. Nelder, J., Mead, R.: A simplex method for function minimization. Comput. J. **7**(4), 308–313 (1965)
21. Ogliari, E., Grimaccia, F., Leva, S., Mussetta, M.: Hybrid predictive models for accurate forecasting in PV systems. Energies **6**(4), 1918–1929 (2013)
22. Reikard, G.: Predicting solar radiation at high resolutions: a comparison of time series forecasts. Sol. Energy **83**(3), 342–349 (2009)
23. Tao, C., Shanxu, D., Changsong, C.: Forecasting power output for grid-connected photovoltaic power system without using solar radiation measurement. In: 2nd IEEE International Symposium on Power Electronics for Distributed Generation Systems (PEDG), HeFei, China, pp. 773–777 (2010)
24. Ulbricht, R., Fischer, U., Lehner, W., Donker, H.: Optimized renewable energy forecasting in local distribution networks. In: Proceedings of the EDBT/ICDT Joint Conference Workshops, EDBT '13, Genua, Italy, pp. 262–266 (2013)
25. Wolff, B., Lorenz, E., Kramer, O.: Statistical learning for short-term photovoltaic power predictions. In: Proceedings of the ECML/PKDD Joint Conference Workshops, Prague, Czech Republik (2013)
26. Yona, A., Senjyu, T., Saber, A.Y., Funabashi, T., Sekine, H., Kim, C.-H.: Application of neural network to 24-hour-ahead generating power forecasting for PV system. In: IEEE Conversion and Delivery of Electrical Energy in the 21st Century, Pittsburgh, USA, pp. 1–6 (2008)
27. Zhang, G.: Time series forecasting using a hybrid ARIMA and neural network model. Neurocomputing **50**, 159–175 (2003)

Forecasting and Visualization of Renewable Energy Technologies Using Keyword Taxonomies

Wei Lee Woon[1]([✉]), Zeyar Aung[1], and Stuart Madnick[2]

[1] Masdar Institute of Science and Technology, P.O. Box 54224, Abu Dhabi, UAE
wwoon@masdar.ac.ae
[2] Massachusetts Institute of Technology, 77 Mass. Ave., Building E53-321,
Cambridge, MA 02139-4307, USA

Abstract. Interest in renewable energy has grown rapidly, driven by widely held concerns about energy sustainability and security. At present, no single mode of renewable energy generation dominates and consideration tends to center on finding optimal combinations of different energy sources and generation technologies. In this context, it is very important that decision makers, investors and other stakeholders are able to keep up to date with the latest developments, comparative advantages and future prospects of the relevant technologies. This paper discusses the application of bibliometrics techniques for forecasting and integrating renewable energy technologies. Bibliometrics is the analysis of textual data, in this case scientific publications, using the statistics and trends in the text rather than the actual content. The proposed framework focuses on a number of important capabilities. Firstly, we are particularly interested in the detection of technologies that are in the *early growth* phase, characterized by rapid increases in the number of relevant publications. Secondly, there is a strong emphasis on visualization rather than just the generation of ranked lists of the various technologies. This is done via the use of automatically generated keyword taxonomies, which increase reliability by allowing the growth potentials of subordinate technologies to be aggregated into the overall potential of larger categories. Finally, by combining the keyword taxonomies with a colour-coding scheme, we obtain a very useful method for visualizing the technology "landscape", allowing for rapidly evolving branches of technology to be easily detected and studied.

1 Introduction

1.1 Motivation

The generation and integration of Renewable Energy is the subject of an increasing amount of research. This trend is largely driven by widely held concerns about the energy sustainability and security and climate change. However, the relevant technical issues are extremely diverse and cover the entire gamut of challenges ranging from the extraction and/or generation of the energy, integration of the energy with existing grid infrastructure and the coordination of energy generation and load profiles via appropriate demand response strategies. For decision

© Springer International Publishing Switzerland 2014
W.L. Woon et al. (Eds.): DARE 2014, LNAI 8817, pp. 122–136, 2014.
DOI: 10.1007/978-3-319-13290-7_10

makers, investors and other stakeholders, the sheer number and variety of the relevant technologies can be overwhelming. In addition this is an area which is evolving rapidly and a huge effort is required just to stay abreast with current development.

All research fields are invariably composed of many subfields and underlying technologies which are related in intricate ways. This composition, or research landscape, is not static as new technologies are constantly developed while existing ones become obsolete, often over very short periods of time. Fields that are presently unrelated may one day become dependent on each others findings. Information regarding past and current research is available from a variety of channels, providing both a difficult challenge as well as a rich source of possibilities. On the one hand, sifting through these databases is time consuming and subjective, while on the other, they provide a rich source of data with which a well-informed and comprehensive research strategy may be formed.

1.2 Theoretical Background

There is already a significant body of research on the topic of technology forecasting, planning and bibliometrics. An in-depth review is beyond the scope of this article but the interested reader is referred to [1–4].

In terms of the methodologies employed, interesting examples include visualizing interrelationships between research topics [5,6], identification of important researchers or research groups [7,8], the study of research performance by country [9,10], the study of collaboration patterns [11–13] and the analysis of future trends and developments [6,14–16]. It is also noteworthy that bibliometric techniques have been deployed on a wide array of research domains, including ones which are related to renewable energies. Some examples include thin film solar cells [17], distributed generation [18], hydrogen energy and fuel cell technology [19,20] and many others.

Our own research efforts have centered on the challenge of *technology forecasting* [21,22], on which this paper is focussed. However, in contrast to the large body of work already present in the literature as indicated above, there is currently very little research which attempts to combine the elements of technology forecasting and visualization.

In response to this apparent shortcoming, in [23] we described a novel framework for automatically visualizing and predicting the future evolution of domains of research. Our framework incorporated the following three key characteristics:

1. A system for automatically creating taxonomies from bibliometric data. We have attempted a number of approaches for achieving this but the basic principle is to create a hierarchical representation of keyword representations where terms that co-occur frequently with one another are assigned to common subtrees of the taxonomy.
2. A set of numerical indicators for identifying technologies of interest. In particular, we are interested in developing a set of simple growth indicators,

similar to technical indicators used in finance. These growth indicators are specially chosen to be easily calculated so that they can be readily applied to hundreds or thousands of candidate technologies at a time. In contrast, traditional curve fitting techniques are more complex and tend to incorporate certain assumptions about the shape in which the growth curve of a technology should take. In addition, more complex growth models require relatively larger quantities of data to properly fit.

3. A technique whereby the afore-mentioned taxonomies can be combined with the growth indicators to incorporate semantic distance information into the technology forecasting process. This is an important step as the individual growth indicators are quite noisy. However, by aggregating growth indicators from semantically related terms spurious components in the data can be averaged out.

In this paper we present further investigations into the use and effectiveness of this framework, particularly in terms of the growth indicators used as well as a more intuitive method of visualizing the scores corresponding to each technology.

2   Analytical Framework

We now describe the framework which will be used to conduct the technology forecasting. However, it is important to first define the form of forecasting that is intended in the present context. It should be pointed out that it is not "forecasting" in the sense of a weather forecast, where specific future outcomes are intended to be predicted with a reasonably high degree of certainty. It is also worth noting that certain tasks remain better suited to human experts; in particular, where a technology of interest has already been identified or is well known, we believe that a traditional review of the literature and of the technical merits of the technology would prove superior to an automated approach.

Instead, the framework proposed in [21] targets the preliminary stages of the research planning exercise by focussing on what computational approaches excel at: i.e. scanning and digesting large collections of data, detecting promising but less obvious trends and bringing these to the attention of a human expert. This overall goal should be borne in mind as, in the following subsections, we present and describe the individual components which constitute the framework.

Figure 1 depicts the high-level organization of the system. As can be seen, the aim is to build a comprehensive technology analysis tool which will collect data, extract relevant terms and statistics, calculate growth indicators and finally integrating these with the keyword taxonomies to produce actionable outcomes. To facilitate discussion, the system has been divided into three segments:

1. Data collection and term extraction (labelled (a) in the figure)
2. Prevalence estimation and calculation of growth indicators (labelled (b))
3. Taxonomy generation and integration with growth indicators (labelled (c))

These components are explained in the following three subsections.

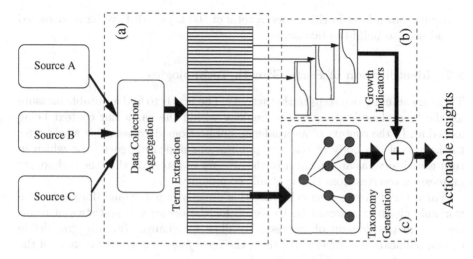

Fig. 1. Analytical framework [23]

2.1   Data Collection and Term Extraction

This consists of the following two stages:

- **Data collection.** The exact collection mechanism, type and number of data sources used are all design parameters that be modified based on user requirements and available resource. However, for the implementation presented in this paper this information was extracted from the Scopus[1] publication database. Scopus is a subscription-based, professionally curated citations database provided by Elsevier. For the results described in this paper, a total of 119,393 document abstracts were collected and processed for subsequent analysis.

  Other possible sources of bibliometrics data that were considered include Google's scholar search engine and ISI's Web of Science database. However, Scopus proved to be a good initial choice as it returned results which were of a generally high quality both in terms of the publications covered and relevance to search terms. In addition, the coverage was sufficiently broad such that searches submitted to Scopus were normally able to retrieve a reasonable number of documents.

- **Term extraction** is the process of automatically generating a list of keywords on which the technology forecasting efforts will be focussed. Again, there are a variety of ways in which this can be achieved; we have experimented with a number of these and our experiences have been thoroughly documented in [24]. For the present demonstration the following simple but effective technique is used: for each document retrieved, a set of relevant keywords is provided. These are collected and, after word-stemming and removal of punctuation marks, sorted according to number of occurrences in the text. For the example

---

[1] http://www.scopus.com

results shown later in this paper, a total of 500 keywords have been extracted and used to build the taxonomy.

2.2   Identification of Early Growth Technologies

There are actually two steps to this activity. The first is to find a suitable measure for the "prevalence" of a given technology within the particular context being looked at. In the context of an academic publications database, this would refer to the size of the body of relevant publications appearing each year which in turn would serve as an indicator of the amount of attention that the technology in question receives from the academic community.

For a variety of reasons achieving this directly is not straightforward but a workable alternative would be to search for the occurrence statistics of terms relevant to the domain of interest. To allow for changes (mainly growth) in overall publication numbers over time, the *term frequency* is used instead of the raw occurrence counts. This is defined as:

$$TF_i = \frac{n_i}{\sum_{j \in \mathcal{I}} n_j} \tag{1}$$

where, n_i is the number of occurrences of keywords i, and $\mathcal{I}$ is the set of terms appearing in all article abstracts (this statistic is calculated for each year of publication to obtain a time-indexed value). Once the term frequencies for all terms have been extracted and saved, they can be used to calculate growth indicators for each of the keywords and hence the associated technologies.

As stated previously, we are most interested in keywords with term frequencies that are relatively low at present but that have been rapidly increasing, which will henceforth be referred to as the "early growth" phase of technological development. Focusing on this stage of technological development is particularly important because we believe that it represents the fields to which an expert would most wish to be alerted since he or she would most likely already be aware of more established research areas while technologies with little observable growth can be deemed to be of lower potential.

Existing techniques are often based on fitting growth curves (see [25] for example) to the data. This can be difficult as the curve-fitting operation can be very sensitive to noise. Also, data collected over a relatively large number of years (approximately ≥ 10 years) is required, whereas the emergence of novel technological trends can occur over much shorter time-scales.

The search for suitable early growth indicators is an ongoing area of research but for this paper we consider the following two indicators as illustrative examples:

$$\eta_i = \frac{[TF_i[y_2] + TF_i[(y_2 + 1)]]}{[TF_i[y_1] + TF_i[(y_1 + 1)]]} \tag{2}$$

$$\theta_i = \frac{\sum_{t \in [y_1, y_2]} t.TF_i[t]}{\sum_{t \in [y_1, y_2]} TF_i[t]}, \tag{3}$$

where, η_i and θ_i is the two different measures of growth for keyword i, $TF_i[t]$ is the term frequency for term i and year t while y_1 and y_2 are the first and last years in the study period.

Hence, η_i gives the ratio of the TF at the end of the study period to the TF at the start of the period, where two year averages are used for the TF terms for improved noise rejection. In contrast, θ_i gives the average publication year for articles appearing over the range of years being studied and which are relevant to term i (a more recent year indicates greater currency of the topic). Using these different expressions provides two separate ways of measuring growth "potential" and helps to avoid confounding effects that may be peculiar to either of these measures.

2.3 Keyword Taxonomies and Semantics Enriched Indicators

One of the problems encountered in earlier experiments involving technology forecasting is that of measuring technology prevalence using term occurrence frequencies. This involves the fundamental problem of inferring an underlying, unobservable property (in this case, the size of the relevant body of literature) using indirect measurements (hit counts generated using a simple keyword search), and cannot be entirely eliminated.

However, one aspect of this problem is a semantic one where individual terms may have two or even more meanings depending on the context. Through our framework an approach was proposed in [23] through which this effect may be reduced. The basic idea is that hit counts associated with a single search term will invariably be unreliable as the contexts in which this term appear will differ. Individual terms may also suffer from the problem of extraneous usage, as in the case of papers which are critical of the technology it represents.

However, if we can find collections of related terms and use aggregate statistics instead of working with individual terms, we might reasonably expect that this problem will be mitigated. We concretize this intuition in the form of a *predictive taxonomy*; i.e. a hierarchical organization of keywords relevant to a particular domain of research, where the growth indicators of terms lower down in the taxonomy contribute to the overall growth potential of higher-up "concepts" or categories.

- **Taxonomy generation** - Taxonomies can sometimes be obtained from external sources and can either be formally curated or "scraped" from online sources such as Wikipedia [26].

  While many of the taxonomies obtained in this way may be helpful for the technology forecasting process, in other cases a suitable taxonomy may simply not be available, or even if available is either not sufficiently updated or is extremely expensive, thus limiting the wider adoption and use of resulting applications. As such, an important capability that has been a focus of our research is the development of a method to perform *automated* creation of keyword taxonomies based on the statistics of term occurrences.

A detailed discussion of this topic is beyond the scope of this paper However, it is sufficient to focus on the basic idea which, as indicated in Sect. 1 is to group together terms which tend to co-occur frequently. Again, we have tested a number of different ways of achieving this (two earlier attempts are described in [23,27] and we have also conducted a survey into different methods of perform taxonomy construction [22]), but in the present context we discuss results produced using one particular method which was found to produce reasonable results while being scalable to large collections of keywords.

This is based on the algorithm described in [28] which was originally intended for social networks where users annotate documents or images with keywords. Each keyword or tag is associated with a vector that contains the annotation frequencies for all documents, and which is then comparable, for e.g. by using the cosine similarity measure. We adapt the algorithm to general taxonomy creation by using two important modifications; firstly, instead of using the cosine similarity function, the *asymmetric* distance function proposed in [23] is used (this is based on the "Google distance" proposed in [29]):

$$\overrightarrow{\mathrm{NGD}}(t_x, t_y) = \frac{\log n_y - \log n_{x,y}}{\log N - \log n_x}, \tag{4}$$

where t_x and t_y are the two terms being considered, and n_x, n_y and $n_{x,y}$ are the occurrence counts for the two terms occurring individually, then together in the same document respectively. Note that the above expression is "asymmetric" in that $\overrightarrow{\mathrm{NGD}}(t_x, t_y)$ refers to the associated cost if t_x is classified as a subclass of t_y, while $\overrightarrow{\mathrm{NGD}}(t_y, t_x)$, corresponds to the inverse relationship between the terms.

The algorithm consists of two stages: the first is to create a similarity graph of keywords, from which a measure of "centrality" is derived for each node. Next, the taxonomy is grown by inserting the keywords in order of decreasing centrality. In this order, each unassigned node, t_i, is attached to one of the existing nodes t_j such that:

$$j = \underset{j \in \mathcal{T}}{\arg\min} \, \overrightarrow{\mathrm{NGD}}(t_i, t_j), \tag{5}$$

(where $\mathcal{T}$ is the set of terms which have already been incorporated into the taxonomy.)

In addition, two further customizable optimizations were added to the basic algorithm described above to improve stability, these are:

1. Attachment of a node to a parent node is based on a weighted average of the similarities to the parent but also to the grandparents and higher ancestors of that node.
2. On some occasions it was necessary to include a "child penalty" whereby the cost of attaching to a given parent increases once the number of children of that parent exceeds a certain number.

These measures and the associated parameters haven't yet been fully explored and in general are set by ad-hoc experimentation. As such they are not discussed in detail in the present context but are the subject of intense further investigations and will be explained in greater detail in future publications.

- **Enhancement and visualization of early growth indicators** - Once the keyword taxonomies have been constructed, they provide a straightforward method of enhancing the early growth indicators using information regarding the co-occurrence statistics of keywords within the document corpus. As with almost all aspects of the proposed framework, a number of variants are possible but the basic idea is to re-calculate the early growth scores for each keyword based on the aggregate scores of each of the keywords contained in the subtree descended from the corresponding node in the taxonomy.

  For the results presented in this paper, aggregation was achieved by simply averaging the respective nodes' scores together with the scores of all child nodes. However, other schemes have also been considered, for example ones which emphasize the score of the current node over the child nodes.

- **Visualization** - The final piece of the puzzle is the development of a good method for representing the results of the above analysis in an intuitive and clear way. A common method for presenting information like this is in the form of a ranked list, which in theory would allow high scoring items to be easily prioritized. However, in practice this can very often produce very misleading results. This is particularly true in our specific application where the target is to study a large numbers of keywords, many of which are closely related. In such a scenario, merely sorting the keywords by their respective scores would most likely result in closely related terms "clumping up" on these lists.

  In contrast, the keyword taxonomy provides a natural alternative framework for achieving this. Firstly, the taxonomy itself allows for relations between the different taxonomies to be easily and quickly grasped. For the growth potentials, we have been experimenting with different ways of representing this information directly within the taxonomies. One simple way is to use a colour-coding scheme where "hot" technologies are coded red (for instance), and there is a range of colours leading up to "cold" technologies in blue. This, in combination with the keyword smoothing step from above means that actively growing areas of researching should turn up as red patches within the taxonomies which can then be detected easily and quickly.

2.4   Renewable Energy Case Study

While this framework can potentially be used on any research domain, we conduct a pilot study on the field of renewable energy to provide a suitable example on which to conduct our experiments and to anchor our discussions. The incredible diversity of renewable energy research as well as the currency and societal importance of this area of research makes it a rich and challenging problem domain on which we can test our methods. Besides high-profile topics like solar cells and nuclear energy, renewable energy related research is also being conducted in fields like molecular genetics and nanotechnology.

To collect the data for use in this pilot study, a variety of high-level keywords related to renewable energy (listed in Sect. 3.1) were submitted to Scopus, and the abstracts of the retrieved documents were collected and used. In total, 119, 393 abstracts were retrieved and subsequently ordered by year of publication.

A number of discussions were held with subject matter experts to identify domains which were both topically current and of high value in the renewable energy industry. The selected domains were Photovoltaics (Solar Panels), Distributed Generation, Geothermal, Wind Energy and Biofuels. Search terms corresponding to each of these domains were then collected and submitted to Scopus' online search interface. These were:

Renewable energy	Embedded generation
Biodiesel	Decentralized generation
Biofuel	Decentralized energy
Photovoltaic	Distributed energy
Solar cell	On-site generation
Distributed generation	Geothermal
Dispersed generation	Wind power
Distrubted resources	Wind energy

3   Results and Discussions

We present results for the renewable energy case study. As described in Sect. 2.1, the Scopus database was used to collect a total of 500 keywords which were relevant to the renewable energy domain, along with 119, 393 document abstracts. These keywords were then used to construct a taxonomy as described in Sect. 2.3, and the growth scores η and θ for each keyword was calculated as shown in Eqs. (2) and (3) respectively.

The two sets of scores thus produced are ranked and the top 30 items from each are presented in Table 1. These scores were subsequently subjected to the taxonomy based aggregation procedure described in Sect. 2.3, producing two further ranked lists which are then presented in Table 2.

Based on these results, some observations are:

1. There were significant differences between the scores obtained using the different growth scoring systems, as well as with and without aggregation, as can be seen from the top-30 lists in Tables 1 and 2. However, at the same time there were also broad similarities between the two sets of rankings which pointed to the underlying "ground truth" which these rankings target as evidenced by a large number of keywords which appeared the top ten items in both lists.

Table 1. Top 30 renewable energy related technology keywords, based on (left) growth ratio (right) average publication year (raw scores)

Growth ratio (η)	Average publication year (θ)
1. Cytology	1. Cytology
2. Biological materials	2. Biological materials
3. Nonmetal	3. Nonmetal
4. Leakage (fluid)	4. Solar equipment
5. Solar equipment	5. Semiconducting zinc compounds
6. Semiconducting zinc compounds	6. Leakage (fluid)
7. Direct energy conversion	7. Direct energy conversion
8. Hydraulic machinery	8. Potential energy
9. Hydraulic motor	9. Alga
10. Potential energy	10. Hydraulic machinery
11. Alga	11. Hydraulic motor
12. Computer networks	12. Ecosystems
13. Bioreactors	13. Bioelectric energy sources
14. Ecosystems	14. Solar power plants
15. Bioelectric energy sources	15. Soil
16. Solar power plants	16. Bioreactors
17. Soil	17. Concentration process
18. Metabolism	18. Solar power generation
19. Concentration process	19. Metabolism
20. Solar power generation	20. Wastewater
21. Wastewater	21. Sugars
22. Sugars	22. Computer networks
23. Nonhuman	23. Nonhuman
24. Experimental studies	24. Experimental studies
25. Zea mays	25. Organic compounds
26. Cellulose	26. Priority journal
27. Priority journal	27. Biomass
28. Organic compounds	28. Lignin
29. Biomass	29. Zea mays
30. Lignin	30. Cellulose

2. In fact, for the aggregated scores, the top six items on both lists are the same (though there were slight differences in the orderings of the terms within this top six set). These were: *cytology, nonmetal, semiconducting zinc compounds, hydraulic machinery, hydraulic motor, alga.* It is interesting to note that these

Table 2. Top 30 renewable energy related technology keywords, based on (left) growth ratio (right) average publication year (with aggregation)

Growth ratio (η)	Average publication year (θ)
1. Cytology	1. Cytology
2. Nonmetal	2. Nonmetal
3. Semiconducting zinc compounds	3. Semiconducting zinc compounds
4. Hydraulic machinery	4. Alga
5. Hydraulic motor	5. Hydraulic machinery
6. Alga	6. Hydraulic motor
7. Direct energy conversion	7. Bioreactors
8. Computer networks	8. Concentration process
9. Solar equipment	9. Metabolism
10. Bioreactors	10. Sugars
11. Cell	11. Computer networks
12. Biological materials	12. Experimental studies
13. Metabolism	13. Ecosystems
14. Concentration process	14. Direct energy conversion
15. Zinc oxides	15. Lignin
16. Potential energy	16. Zea mays
17. Sugars	17. Bioelectric energy sources
18. Ecosystems	18. Phosphorus
19. Bioelectric energy sources	19. Biological materials
20. Experimental studies	20. Cellulose
21. Zea mays	21. Nitrogenation
22. Soil	22. Bacteria (microorganisms)
23. Cellulose	23. Adsorption
24. Lignin	24. Soil
25. Hydrolysis	25. Hydrolysis
26. Photovoltaic cell	26. Glycerol
27. Fermenter	27. Fermenter
28. Glucose	28. Glucose
29. Glycerol	29. Potential energy
30. Adsorption	30. Biodegradable

terms correspond to important research topics within three separate sub-domains of renewable energy - biomass, solar cells and wind power.

3. It was interesting to note the number of biotechnology related keywords that were found in all four lists. This reflects the fact that biological aspects of renewable energy are amongst the most rapidly growing areas of research.

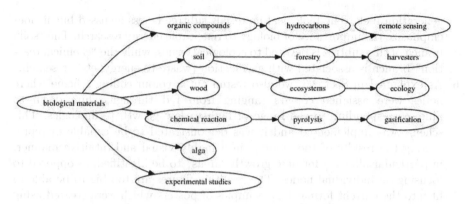

Fig. 2. Subtree for node "Biological materials"

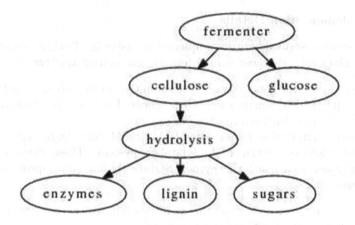

Fig. 3. Subtree for node "fermenter"

Amongst the highly-rated non-biological terms on the list were "nonmetal" (#2) and "seminconducting zinc compounds" (#3), both of which are related to the field of thin-film photovoltaics.

4. However, many of the keywords in the lists in Tables 1 and 2 were associated with leaves in the taxonomy; this was a desirable outcome, as these were the less well known and hence more interesting technologies, but it also meant that the confidence in the scores were lower. Looking at the terms with relatively large associated subtrees, we see that three of the largest were "biological materials" (15 nodes), "fermenter" (7 nodes) and "hydrolysis" (4 nodes). The subtrees for the first two terms are shown in Figs. 2 and 3 respectively, while the hydrolysis subtree is actually part of the "fermenter" subtree and as such is not displayed.

5. The fermenter subtree is clearly devoted to biofuel related technologies (in fact, two major categories of these technologies are represented - "glucose"-related or first generation biofuels, and "cellulosic" biofuels which are second

generation fuels. The biological materials subtree is less focussed but it does emphasize the importance of biology to renewable energy research. The "soil" branch of this subtree is devoted to ecological issues, while the "chemical reaction" branch is associated with gasification (waste-to-energy, etc.) research.

6. As explained in Sect. 2.3, we also tested out a colour-coding scheme where nodes were assigned colours ranging from red through white down to blue, corresponding to the range of high to low growth technologies. This scheme was implemented and it was demonstrated to be capable of representing the results of the analysis in a highly visual and intuitive manner, in particular allowing for high growth "areas" to be identified, as opposed to focusing on individual nodes. The resulting figures are too big to be able to fit into the current format but examples of posters which were created using this technique can be viewed at: http://www.dnagroup.org/posters/.

3.1   Implementation Details

The framework described here was implemented using the Python programming language. Data collection was semi-automatic and consisted of two main steps:

1. Customized keyword queries were first submitted to the Scopus search portal. The results of these searches were then exported as comma-delimited (*.csv) files and downloaded from the Scopus website.
2. Automated scripts were then created to filter and store the records in a local SQL database (we used the SQLite database system). These which were subsequently accessed using the python SQLite toolkit and appropriate SQL language calls.

Figures were generated using the *pydot* toolkit which provides a Python based interface to the Graphviz Dot language[2].

4   Conclusion

In this paper, we present the use of an innovative framework for visualizing the research "landscape" of the domain of renewable energy. Sucha framework will be extremely useful for supporting the relevant research planning and decision-making processes.

The system covers the entire chain of activities starting with the collection of data from generic information sources (online or otherwise), the extraction of keywords of interest from these sources and finally the calculation of semantically-enhanced "early growth indicators". Finally, a colour-coding scheme is used to annotate the resulting taxonomies, allowing rapidly growing areas of research to be easily detected within the overall context of the research domain.

The simple implementation of this framework presented in this paper is used to study developments within the domain of renewable energy. More analysis is

---

[2] http://code.google.com/p/pydot/

required before deeper insights can be gained and these results can be applied "on the field" by investors and other stakeholders. However, we note that the results of the analysis do seem to reflect factors and developments within the field of renewable energy.

The results of this effort are presented and discussed. While the current implementation still has ample scope for future extensions, the results are already encouraging though currently the process is still a little too noisy to pick out "very early growth" technologies. However, we are investigating numerous avenues for enhancing the basic implementation referenced here, and are confident of presenting improved findings in upcoming publications.

References

1. Porter, A.L.: Technology foresight: types and methods. Int. J. Foresight Innov. Policy **6**(1), 36–45 (2010)
2. Porter, A.L.: How "tech mining" can enhance R&D management. IEEE Eng. Manag. Rev. **36**(3), 72 (2008)
3. Martino, J.P.: A review of selected recent advances in technological forecasting. Technol. Forecast. Soc. **70**(8), 719–733 (2003)
4. Martino, J.: Technological Forecasting for Decision Making. McGraw-Hill Engineering and Technology Management Series. McGraw-Hill, New York (1993)
5. Porter, A.: Tech mining. Compet. Intel. Mag. **8**(1), 30–36 (2005)
6. Small, H.: Tracking and predicting growth areas in science. Scientometrics **68**(3), 595–610 (2006)
7. Kostoff, R.N., Toothman, D.R., Eberhart, H.J., Humenik, J.A.: Text mining using database tomography and bibliometrics: a review. Technol. Forecast. Soc. Chang. **68**(3), 223–253 (2001)
8. Losiewicz, P., Oard, D., Kostoff, R.: Textual data mining to support science and technology management. J. Intell. Inf. Syst. **15**(2), 99–119 (2000)
9. de Miranda Santo, M., Coelho, G.M., dos Santos, D.M., Filho, L.F.: Text mining as a valuable tool in foresight exercises: a study on nanotechnology. Technol. Forecast. Soc. **73**(8), 1013–1027 (2006)
10. Kim, M.-J.: A bibliometric analysis of the effectiveness of koreas biotechnology stimulation plans, with a comparison with four other asian nations. Scientometrics **72**(3), 371–388 (2007)
11. Anuradha, K., Urs, S.R.: Bibliometric indicators of indian research collaboration patterns: a correspondence analysis. Scientometrics **71**(2), 179–189 (2007)
12. Chiu, W.-T., Ho, Y.-S.: Bibliometric analysis of Tsunami research. Scientometrics **73**(1), 3–17 (2007)
13. Braun, T., Schubert, A.P., Kostoff, R.N.: Growth and trends of fullerene research as reflected in its journal literature. Chem. Rev. **100**(1), 23–38 (2000)
14. Smalheiser, N.R.: Predicting emerging technologies with the aid of text-based data mining: the micro approach. Technovation **21**(10), 689–693 (2001)
15. Daim, T.U., Rueda, G.R., Martin, H.T.: Technology forecasting using bibliometric analysis and system dynamics. In: Technology Management: A Unifying Discipline for Melting the Boundaries, pp. 112–122 (2005)
16. Daim, T.U., Rueda, G., Martin, H., Gerdsri, P.: Forecasting emerging technologies: use of bibliometrics and patent analysis. Technol. Forecast. Soc. **73**(8), 981–1012 (2006)

17. Guo, Y., Huang, L., Porter, A.L.: The research profiling method applied to nano-enhanced, thin-film solar cells. R&D Manag. **40**(2), 195–208 (2010)
18. Woon, W.L., Zeineldin, H., Madnick, S.: Bibliometric analysis of distributed generation. Technol. Forecast. Soc. Chang. **78**(3), 408–420 (2011)
19. Tsay, M.-Y.: A bibliometric analysis of hydrogen energy literature, 1965–2005. Scientometrics **75**(3), 421–438 (2008)
20. Yu-Heng, C., Chia-Yon, C., Shun-Chung, L.: Technology forecasting of new clean energy: the example of hydrogen energy and fuel cell. Afr. J. Bus. Manag. **4**(7), 1372–1380 (2010)
21. Woon, W.L., Henschel, A., Madnick, S.: A framework for technology forecasting and visualization. In: International Conference on Innovations in Information Technology, 2009. IIT'09, pp. 155–159. IEEE (2009)
22. Henschel, A., Woon, W.L., Wachter, T., Madnick, S.: Comparison of generality based algorithm variants for automatic taxonomy generation. In: International Conference on Innovations in Information Technology, 2009. IIT'09, pp. 160–164. IEEE (2009)
23. Woon, W.L., Madnick, S.: Asymmetric information distances for automated taxonomy construction. Knowl. Inf. Syst. **21**(1), 91–111 (2009)
24. Ziegler, B., Firat, A.K., Li, C., Madnick, S., Woon, W.L.: Preliminary report on early growth technology analysis. Technical report CISL #2009-04, MIT (2009). http://web.mit.edu/smadnick/www/wp/2009-04.pdf
25. Bengisu, M., Nekhili, R.: Forecasting emerging technologies with the aid of science and technology databases. Technol. Forecast. Soc. **73**(7), 835–844 (2006)
26. Dawelbait, G., Henschel, A., Mezher, T., Woon, W.L.: Forecasting renewable energy technologies in desalination and power generation using taxonomies. Int. J. Soc. Ecol. Sustainable Dev. (IJSESD) **2**(3), 79–93 (2011)
27. Woon, W.L., Madnick, S.: Semantic distances for technology landscape visualization. J. Intell. Inf. Syst. **39**(1), 29–58 (2012)
28. Heymann, P., Garcia-Molina, H.: Collaborative creation of communal hierarchical taxonomies in social tagging systems. Technical report, Stanford University, #2006-10 (2006). http://dbpubs.stanford.edu:8090/pub/2006-10
29. Cilibrasi, R.L., Vitányi, P.M.B.: The google similarity distance. IEEE Trans. Knowl. Data Eng. **19**(3), 370–383 (2007)

Rooftop Detection for Planning of Solar PV Deployment: A Case Study in Abu Dhabi

Bikash Joshi, Baluyan Hayk, Amer Al-Hinai, and Wei Lee Woon$^{(\boxtimes)}$

Masdar Institute of Science and Technology, P.O. Box 54224,
Abu Dhabi, UAE
wwoon@masdar.ac.ae

Abstract. Photovoltaic (PV) technology is one of two modes of energy generation which utilize solar energy as its source. The rooftops of buildings can be utilized for solar power generation using this technology, and are considered to be highly promising sites for urban PV installations due to land space limitations. However, to properly plan such installations decision makers would need to have detailed information about the amount of rooftop area that is available, as well as the distribution of individual rooftop sizes. In this paper a machine learning based approach for detecting rooftops is proposed and its utility for planning rooftop PV installations is demonstrated via a simple pilot study on two different residential areas in Abu Dhabi, UAE. The proposed method uses a two-stage classification model to estimate the rooftop area that is available for solar panel installation. Next, a comparative study of three different types of PV technologies is conducted in terms of energy generation and economic viability. The results obtained from these experiments suggest that thin-film panels may have a distinct advantage over other PV technologies. Even though the cost of PV panels is still quite high, this could be balanced by the potential benefits to the environment. If reasonable subsidies and feed-in tariffs are implemented, PV technology can become a cost effective option for the UAE.

Keywords: Rooftop detection · PV systems · Energy generation · Economic analysis

1 Introduction

1.1 Background and Motivation

The World Energy Council has estimated that the earth's surface receives around 3,850,000 EJ (exajoules; $1\,EJ \approx 10^8\,J$) of solar energy annually; this translates to the annual global energy consumption in 2002 being received in an hour. However, in 2007 only 0.1 % of the world's total energy consumption is fulfilled by solar energy, indicating that solar energy is still greatly under-utilized. However, more recent reports show that the cumulative installed capacity of solar PV has exceeded 35000 MW in IEA member countries, up from 500 MW at the end

© Springer International Publishing Switzerland 2014
W.L. Woon et al. (Eds.): DARE 2014, LNAI 8817, pp. 137–149, 2014.
DOI: 10.1007/978-3-319-13290-7_11

of 1994. This points to the increasing popularity of PV systems as a source of energy. Also, recent studies have shown that since 1994 the installation of solar PV systems has increased at an annual average of greater than 25 % [1].

The United Arab Emirates (UAE) is rich in conventional energy resources i.e. petroleum and natural gas. 9.3 % of the worlds oil reserves and 4.1 % of worlds gas reserves are in the UAE. Abu Dhabi, the capital of the UAE and dubbed the "richest city in the world" by CNN, contains the majority of the UAE's fossil fuel reserves, with 95 % of oil and 92 % of gas deposits. However, Abu Dhabi is already planning ahead and has begun to invest heavily in renewable energy technologies (RETs). In this way it hopes to maintain its position as a global leader in energy well into the post-fossil fuel era [2].

With a rapidly growing population and urbanization, there is a continuous need for energy. The opportunity for renewable energy sources to become popular has been driven by this increasing demand for energy. In last three decades, there was an exponential increase in the electricity consumption in the UAE. Also, this demand for electricity in Abu Dhabi is expected to increase four fold from 5616 MW in 2008 to 25,530 MW in 2028. Furthermore, as a wealthy city in UAE, Abu Dhabi can afford to have solar energy projects, which requires high investment [2].

By announcing the goal of generating 7 % of its power from renewable energy sources by the year 2020, the UAE has plotted an ambitious route towards the national uptake of renewable energies. The number of rooftop PV stations in the UAE is expected to increase significantly in the near future and could include installations on both residential or commercial buildings. The proper use of rooftop space for PV deployment can help to avoid potential land use. More benefits offered by rooftops for deployment of renewable solar energy include access to sunlight, low-cost solar real estate, and attractive investment. However, to properly plan such installations decision makers would need to have detailed information about the amount of rooftop area that is available, as well as the distribution of individual rooftop sizes. One way to address this requirement is via the use of computational methods of rooftop detection, which will be discussed next.

1.2 Rooftop Detection

Computational solutions for rooftop detection are based on image processing operations such as edge detection, corner detection and image segmentation. The basic approach is to first generate rooftop candidates using image segmentation techniques and then to identify true rooftops using features like shape, area and the presence of shadow [3,4]. Machine learning methods have proved particularly popular in recent studies, where Artificial Neural Networks (ANN) and Support Vector Machines (SVM) have been widely used.

In [5] an ANN is used to facilitate effective rooftop identification in the presence of noise and artifacts. In [6] a review is presented of ANN based approaches in various related areas including shape segmentation in medical images, biometric patterns and gestures extraction, letters and characters detection, edge and contour detection. For each of these applications ANNs performed reasonably well, thus demonstrating its potential for solving image processing tasks. Neural network based methods have also been used to segment images into parts that meet certain criteria [7]. Most of these approaches are based on pixel-level segmentation, which assign each pixel to a given segment based on features generated for the pixel in question. However this approach does not perform well for rooftop detection; in particular these methods fail to detect rooftops when applied to test images containing objects (e.g. cars, roads) which are of the same color as rooftops.

To address these apparent shortcomings, we use an innovative method which is based on two consecutive classification stages [8,9]. A Multi-Layer Perceptron (MLP) neural-network is first used to provisionally classify candidate segments as either rooftop or non-rooftop. New features are then extracted from the outputs of the MLP which are based on rooftop properties inferred using results gathered over the entire image. The second classification stage is then performed on these features using an SVM. While this is not unique, we also note that there appear to be relatively few studies in which SVMs are used for object detection (notable examples are [10] and [11]). Dividing the classification process into two separate stages appears to reduce the number of false positives, resulting in a significant improvement in the performance of the model in comparison with traditional approaches.

1.3 Objectives

The overall goal of our work is to assess the potential of solar PV energy production from residential rooftops in Abu Dhabi, UAE. This is achieved via two key stages.

Firstly, we describe an innovative dual-stage rooftop detection system which accurately estimates the location and size of rooftops within satellite images. This information is then used to evaluate the viability of solar PV installations based on the following two aspects:

- **Technical analysis of solar PV system:** We first estimate the total annual energy generated and peak power of the solar PV panels when installed on the rooftops of residential buildings.
- **Economic analysis of the system** in terms of economic indicators such as NPV, IRR, payback period etc. for different types of solar PV panels.

We intend to evaluate each of these aspects for the three different types of solar PV panels (Monocrystalline, Polycrystalline and Thin-film panels) and present comparative analysis of different PV panels.

2   Methodology

The proposed methodology consists of three main steps: image segmentation, first- then second-stage classification. Each of these steps will now be discussed in greater detail:

2.1   Image Segmentation

Bilateral filtering is first used to enhance the image, which is then divided into a set of segments using k-means clustering. Each of these segments is considered as a candidate for being a rooftop or part of rooftop. The clustering of the color images is performed in the RGB intensity space.

By trying different values of k, it was determined that $k = 4$ gave the best segmentation result for our images. After all the pixels had been divided into 4 clusters, candidate regions were generated by finding connected-regions - i.e. regions where pixels of the same cluster were adjacent to each other. The 4-connected flood fill algorithm was used for this purpose.

2.2   First Classification Stage

A set of 15 training images were prepared by manually labelling rooftops present in these images. Each of these training images were then segmented into regions from which a set of 14 features were extracted. Features are numerical attributes which allow rooftop and non-rooftop regions to be distinguished from each other. In this study, 14 features which were relevant to the classification task at hand were selected. These are: (1) Area (2) Minor to major axis ratio (3) Mean intensity (4) Solidity (5) Variance (6) Entropy (7) Roundness (8) Rectangularity (9) Contrast (10) Correlation (11) Homogeneity (12) Number of Corners (13) Energy. Each row in the dataset hence corresponds to one image segment and is manually labeled as "1" (if it corresponds to a rooftop) or "0" (if not).

The training dataset then is then used to train an MLP, which is then presented with each test image. The output will then be the predicted labels and class probabilities for each candidate region.

2.3   Second Classification Stage

In practice, the MLP will not be able to detect all of the rooftop regions in the image. In particular we noticed the presence of many false positives in this first pass of labeled regions. To address this problem, a second classification stage is performed which uses information from the results of the first classification stage. This also helps to identify some of the rooftops which were missed in the first-stage classification. The two main components of this stage are:

– **Feature "Re-extraction"**
  The following features are used for the second stage classification:
  1. *Class Probability:* This is the output of the MLP for each segment of the test image.
  2. *Confidence:* Rooftops in one region are similar to each other. So, the result of first-stage classification shows the most prevalent intensity value of rooftop. A histogram is used to characterize the intensity values of the rooftops identified by first-stage classification. The intensity range is divided into 8 bin, and the number of pixels falling into each bin provides an associated confidence value. During the second classification stage this confidence value is then used as a feature.

3. *Shadow:* We also use building shadows as a feature for the second-stage classification. It is obvious that a building rooftop region should be accompanied by its shadow.

In particular, note that the second item in the list above serves to incorporate information from the entire image into the classification of individual regions. It was felt that this was of particular importance in improving the performance of our method.

- **Classification using Support Vector Machine(SVM)**
  SVM is a supervised learning technique which finds an optimal decision boundary to separate the feature space into the corresponding classes. One important consideration during the use of SVM is the choice of kernel function. Preliminary investigations found that the polynomial kernel function gave the best results and this was used for all subsequent experiments.

  First, we trained a SVM using the training images. Then we test each of our test images using the trained SVM model. Thus, we get our final rooftop and non-rooftop labels.

2.4 Estimation of Energy Generation

The solar energy incident on a surface for a particular period of time depends on the solar insolation, and is calculated by the product of insolation level, the surface area under study and duration of exposure. The amount of energy that can subsequently be extracted as electricity is further dependent on the efficiency of the PV panel used as well as its surface inclination, irregularities and dust particles. These accumulated losses are accounted for by a factor known as the *performance ratio*, for which a reasonable estimate for many situations is 0.75[1]. The solar radiation can be direct radiation, diffused radiation or reflected radiation. We have used the Global Solar Radiation (GSR) which accounts for all these forms of radiation and is measured in terms of $kWh/m^2/day$. The GSR data is estimated by NASA by averaging the 22 years of study is used from [2]. The highest intensity of radiation is received in May and June, at $7.6\,kWh/m^2/day$ while the minimum intensity is in December at $4\,kWh/m^2/day$. The energy converted by the solar panel can be estimated as:

$$E_{gen} = \eta \times GSR \times A_{eff} \times N_{days} \times PR$$

$$Where,$$

$$E_{gen} = \text{Energy generated in kWh}$$

$$\eta = \text{Efficiency of solar cell}$$

$$GSR = \text{Global Solar Radiation}$$

$$N = \text{Number of days}$$

$$A_{eff} = \text{Effective rooftop area}$$

$$PR = \text{Performance Ratio}$$

(1)

---

[1] http://photovoltaic-software.com/PV-solar-energy-calculation.php

These statistics can be used to calculate the annual energy generation by summing the energy generation for all the months. As per [12], 65 % of the aggregate rooftop area is actually available for solar PV installation, which is referred to as effective rooftop area. Hence, the utilization factor for area is taken as 0.65. This effective rooftop area is used in all the subsequent calculations. Performance Ratio (PR) is used to evaluate the quality of PV installation as it gives the performance of the installation independent of the orientation, inclination of the panel. It includes all the losses in the PV system.

The PV panel capacity that can be installed in the effective roof area is calculated using Eq. 2.

$$\text{Potential Power} = \frac{\text{Annual Energy Generation}}{\text{Annual Peak Sunshine Hour}} \tag{2}$$

Here, the peak sunshine hours refers to the average sunshine hours that has $1\,kW/m^2$ radiation. The peak sunshine for Abu Dhabi is 2179.5 h/year [2].

2.5 Economic Analysis of PV System

The main objective of economic analysis of PV power plant for three different PV technologies is to draw out the most feasible technology for rooftop. It also takes into account of the elements that effect the profitability of the project. The project life is assumed to be 25 years [2]. We have taken into account the following transactions for the economic analysis:

1. Upfront Cost: The total upfront cost of solar PV system installation is estimated by summing up the cost of installation of individual components of solar system, which in turn is primarily determined by the cost of the Inverter and the Solar Panels themselves[2]. These are estimated as follows:
   (a) Cost of inverter: We know that, power is the rate at which energy is generated. Power is specified in terms of kW (kilowatts) for electrical appliances. So, the overall cost of the inverters can be calculated as:

$$Cost_{inverter} = P_{peak,inverter} \times Cost_{inverter,kW} \\ \times \text{Number of inverters} \tag{3}$$

Where, $Cost_{inverter,kW}$ is cost of inverter per kilowatt and $P_{peak,inverter}$ is the peak power of the inverter.

   (b) Cost of solar panels: The total cost of solar panels can be calculated using the below formula:

$$Cost_{panels} = P_{peak,panels} \times Cost_{panels,kW} \tag{4}$$
$$Where,$$
$$P_{peak,panels} = \text{Total peak power of panels}$$
$$Cost_{panels} = \text{Total cost of panels}$$
$$Cost_{panels,kW} = \text{Cost of panels per kW}$$

The panel cost per W for different PV panels are shown in Table 5.

---

[2] http://www.nmsea.org/Curriculum/7_12/Cost/calculate_solar_cost.htm

2. Operation and Maintenance (O&M) Cost: This is the annual cost of operating and maintaining the system. It is assumed to be fixed throughout the lifetime of the project, and is fixed at an estimated level of 33.5 $/kW.

3. Inverter Replacement (IR) Cost: Inverters have finite operating lifetimes, which depend on the specific type and manufacturer. Here, an average lifetime of 5 years is assumed [2]. This sets up a replacement cost which recurs every 5 years until the end of the project.

4. Revenue from Electricity Generation: This is the main income from the project. We assume that electricity generated from the PV installations will be sold directly to the electric grid at the prevailing feed-in tariff; for the UAE a figure of 0.0816 is used [2].

5. Salvage Value: At the end of the project, the components of the project will still have monetary value, which is known as salvage value. The salvage value is usually assumed to be 10 % of the initial investment [2].

2.6   Economic Indices

The following indices are used to evaluate the economic feasibility of the different PV technologies:

1. Net Present Value (NPV): NPV is the difference between the present value of all cash inflows and the present value of all cash outflows, where the "present value" of a cash flow is the amount of this flow discounted to its present value as shown in Eq. (5). The net present value of an economically feasible project should be positive [13].

$$F = P(1 + i)^N,$$ (5)

where F =Future Value, P =Present Value, i =Discount Rate, N =time.

2. Internal Rate of Return (IRR): The Internal Rate of Return measures the compound rate at which benefit is made from a project. It is the rate at which NPV becomes zero. IRR of any project can be estimated by using the NPV expression shown in Eq. 5 and equating it to zero. This equation is non linear and has many solutions. Out of multiple solutions, the required IRR should be positive [14].

3. Simple Payback Period (SPP): The period of time in years at which a project earns all of its investment and after which it starts making profit. Lower payback period signifies a higher project feasibility. Simple Payback Period is calculated using Eq. 6.

$$SimplePaybackPeriod = \frac{InitialInvestmentCost}{AnnualOperatingSavings}$$ (6)

4. Benefit Cost Ratio (BCR): Benefit Cost Ratio compares the profit from the project with the costs involved. The revenues and profits are taken as benefits whereas the investment and expenses as taken as positive cost. The salvage value is taken as negative cost. Finally, benefit by cost ratio can be expressed as shown in Eq. 7.

$$BenefitCostRatio(BCR) = \frac{PVofrevenue}{PVofcost - PVofsalvage} \qquad (7)$$

3   Result and Discussion

3.1   Rooftop Area Calculation

Images depicting residential areas of Abu Dhabi were used. In order to check the generality of proposed model, images from two separate areas with different characteristics were used. The two image datasets were named the Khalifa and Raha datasets corresponding to the names of the two residential areas.

To facilitate better image segmentation, the images were divided into 512×512 pixel sized tiles, which corresponds to approximately 70×70 square metres of land area. Both Khalifa and Raha dataset consist of 25 images each of which 15 images were used for training and 10 were used for testing.

Classification Precision and Recall are used to evaluate the performance of the system, defined as:

$$\text{Precision} = \frac{\text{TP}}{\text{TP} + \text{FP}} \qquad \text{Recall} = \frac{\text{TP}}{\text{TP} + \text{FN}}$$

Where TP, FP, TN and FN are True Positive, False Positive, True Negative and False Negative rates respectively.

Sample result of our algorithm is shown in Fig. 1 It can be seen that the result of first-stage classification contains many false positive regions. This is what our second-stage classification helps to improve. Second-stage classification reduces false positives largely, which in turn significantly improves the precision. Even though there is slight decrease in recall values as some of the rooftops identified in the first-stage may be lost, the improvement in precision and false positives is significant.

The overall performance of our method in the "Khalifa" and "Raha" datasets is shown Table 1. This shows the overall precision, recall and FP values for the test datasets.

As explained in the methodology section, aggregate rooftop area is estimated using the two-stage classification procedure. 65 % of the aggregate rooftop area is available for solar PV installation, which is referred to as effective rooftop area. The calculated values of aggregate rooftop area and effective rooftop area for both the datasets are listed in Table 2.

3.2   Annual Solar Energy Generation

Equations 1 and 2 are then used to estimate the solar energy generation and solar peak power for three types of solar panels. The parameters used for this calculation are listed in Table 3. The calculated values of annual energy generation and peak power are shown in Table 4. the chart, the energy generation from thin-film cells is significantly less than that for the other two types of solar panels. Whereas the energy generation from monocrystalline solar panels is slightly higher than that for polycrystalline panels.

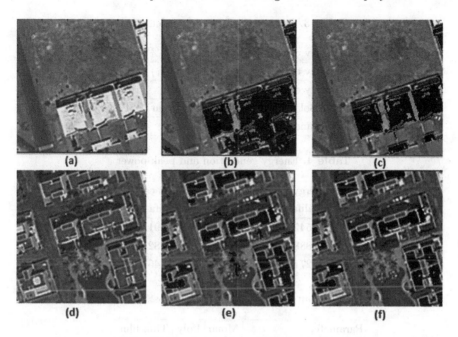

Fig. 1. The original image from Khalifa City A (a) The result after first-stage classification (b) The result after the second-stage classification (c) The original image from Raha Gardens (d) The result after first-stage classification (e) The result after the second-stage classification (f)

Table 1. Results using Khalifa and Raha test dataset

	First-stage			Second-stage		
	Precision	Recall	FP	Precision	Recall	FP
Khalifa	89.95	82.39	9.21	92.24	81.16	6.82
Raha	87.78	86.69	12.16	92.29	83.08	7.07

Table 2. Area calculation for Khalifa and Raha datasets

	Khalifa	Raha
Actual aggregate area (m^2)	13929.61	6119.83
Detected aggregate area (m^2)	13958.06	6304.23
Effective area (m^2)	9072.74	4097.75

3.3 Upfront Cost Calculation

Following the approach presented in methodology section, we calculated the upfront cost of the PV system. Various data such as market price,

Table 3. Parameters for energy calculation

Efficiency of monocrystalline panels (%)	20
Efficiency of polycrystalline panels (%)	18
Efficiency of thin-film panels (%)	10
Global Solar Radiation (kWh/m^2/year)	2179
Performance ratio assuming losses	0.75

Table 4. Energy generation and peak power

Panel	Energy generation (kWh)		Plant capacity (kW)	
	Khalifa	Raha	Khalifa	Raha
Mono	2965424.74	1339349.42	1360.91	614.66
Poly	2668882.27	1205414.48	1224.82	553.19
Thin-film	1482712.37	669674.71	680.46	307.33

Table 5. Input parameters for upfront cost calculation

Parameter	Mono	Poly	Thin-film
$P_{peak,inverter}(kW)$	3	3	3
$Cost_{inverter,kW}(\$)$	200	200	200
$Cost_{panel,\$/kW}$	3000	2000	1000
$Roof Area_{Khalifa}(m^2)$	150	150	150
$Roof Area_{Raha}(m^2)$	70	70	70

Table 6. Total upfront cost for Khalifa and Raha datasets

Solar panel type	Upfront cost (\$)	
	Khalifa	Raha
Monocrystalline	4119023.51	1879110.84
Polycrystalline	2485930.49	1141515.93
Thin-film	716746.38	342454.78

specification et al. obtained from [15][3,4] are used for this calculation. These parameters used for this calculation are listed in Table 5. Table 6 presents the calculated values of upfront costs for three types of solar panels. As expected, the upfront cost of monocrystalline solar panels is maximum, whereas it is slightly less for polycrystalline panels and it is lowest for thin-film panels.

---

[3] http://grensolar.com/solar_products/solar_panel/

[4] http://www.nmsea.org/Curriculum/7_12/Cost/calculate_solar_cost.htm

3.4   Economic Analysis

Data obtained from [2,13,15–17] and the results presented in the previous sections are then used to analyze the economic viability of the PV system when implemented using different PV panel technologies. The various parameters used in the calculation are listed in Table 7.

Table 7. Parameters for economic analysis

Parameter	Mono	Poly	Thin-film
O&M cost ($/kW)	33.5	33.5	33.5
IR cost Khalifa ($)	36216.99	36216.99	36216.99
IR cost Raha ($)	34096.19	34096.19	34096.19
Discount rate (%)	7.55	7.55	7.55
Lifetime (years)	25	25	25
Salvage value Khalifa($)	411062.79	248086.35	71528.55
Salvage value Raha($)	182414.65	110812.64	33243.79
Feed-in-tariff ($/kWh)	0.0816	0.0816	0.0816
Panel cost ($/Watt)	3	2	1

Using these parameters, we calculated the NPV, IRR, SPP and BCR for three types of panels for both Khalifa and Raha dataset. The results are shown in Tables 8 and 9.

The NPV and BCR for mono and poly crystalline panels are negative and less than unity respectively. These values of NPV and BCR indicate that the project is economically infeasible with mono and polycrystalline panels. In contrast,

Table 8. Economic indices for Khalifa Dataset

Index	Mono	Poly	Thin-film
NPV	−1936091	−547424	321260
IRR (%)	1.71	4.99	12.30
SPP (Years)	20.94	14.23	7.66
BCR	0.58	0.82	1.31

Table 9. Economic indices for Raha dataset

Index	Mono	Poly	Thin-film
NPV	−925595	−298396	93949
IRR (%)	1.38	4.47	10.54
SPP (Years)	21.87	15.03	8.71
BCR	0.57	0.79	1.18

the NPV and BCR for thin-film panels are much more positive. Even though more electricity can be generated using mono and poly crystalline panels, the associated costs are also a lot higher. The IRR value for thin-film panels is 10–12% and its payback period is 7-8 years. Again, these figures are better than those obtained for mono and poly crystalline panels. In addition, similar conclusions were reached when studying the results for both datasets i.e. Khalifa and Raha.

4   Conclusion and Future Work

In this paper, a two-stage classification technique is used to determine the roof area for quantifying rooftop solar PV potential. Images obtained from Google Maps were used in the experiments.

Based on these estimates, three different PV technologies were then evaluated based on the respective energy generation potentials and economic feasibility. The case study was done for building rooftops in two residential regions of Abu Dhabi, UAE.

Abu Dhabi has huge potential for electricity generation using rooftop PV systems. As per the results of this project, the NPV for the monocrystalline and polycrystalline PVs are negative whereas it is positive for thin film panels. This shows that monocrystalline and polycrystalline PVs are unlikely to be profitable even though they generate more energy. One main reason for this is the low feed-in-tariff in the UAE. However, if the feed-in-tariff is increased by some extent, the NPV of monocrystalline and polycrystalline panels can be improved. In contrast, the thin film panels were found to be cheaper and economically feasible for rooftop power generation. The payback period and IRR of thin film was found to be around 7–9 years and 11–13% respectively. Thin film panel was found to have higher IRR, NPV and BCR in comparison to mono and polycrystalline cells.

PV technology provides significant environmental benefits, even though the associated costs are still quite high. The adoption of rooftop solar PV technology will have social and environmental benefits beyond the cost considerations. However, it will be difficult to attract investors for PV installations without a strong economic case and, if required, attractive incentives. The latter could be in the form of subsidy or improved feed-in-tariff. The work presented in this paper can be further extended in the following directions:

1. In this paper we calculated the aggregated rooftop area for two residential regions of Abu Dhabi city using machine learning based technique. Similar study can be performed for a different geographical region and the approach can be tested in a large geographical area.
2. In Abu Dhabi, most of the residential buildings are of same height. However, if the buildings are of uneven heights, the shadow of one building falls on another. In such cases, the portion of shadowed roof should be discarded while calculating energy generation. So, we plan to study the effect of such shadow on energy generation from PV.

3. The carbon saving from PV installation can also be considered in terms of monetary value and taken as a benefit in the economic analysis. This would certainly improve the NPV, IRR, SPP and benefit cost ratio.

References

1. Zahedi, A., Lu, J.: Economic evaluation of grid-connected solar PV production cost in New Zealand. In: 2012 IEEE International Conference on Power System Technology (POWERCON), pp. 1–4. IEEE (2012)
2. Harder, E., MacDonald, J.: The costs and benefits of large-scale solar photovoltaic power production in Abu Dhabi, United Arab Emirates. Renew. Energy **36**(2), 789–796 (2011)
3. Nosrati, M.S., Saeedi, P.: A novel approach for polygonal rooftop detection in satellite/aerial imageries. In: 2009 16th IEEE International Conference on Image Processing (ICIP), pp. 1709–1712 (2009)
4. Izadi, M., Saeedi, P.: Automatic building detection in aerial images using a hierarchical feature based image segmentation. In: 2010 20th International Conference on Pattern Recognition (ICPR), pp. 472–475 (2010)
5. Maloof, M.A., Langley, P., Binford, T.O., Nevatia, R., Sage, S.: Improved rooftop detection in aerial images with machine learning. Mach. Learn. **53**, 157–191 (2003)
6. Ramírez-Quintana, J.A., Chacon-Murguia, M.I., Chacon-Hinojos, J.F.: Artificial neural image processing applications: a survey. Eng. Lett. **20**(1), 68 (2012)
7. Egmont-Petersen, M., de Ridder, D., Handels, H.: Image processing with neural networks a review. Pattern Recogn. **35**(10), 2279–2301 (2002)
8. Joshi, B., Baluyan, H., Al-Hinai, A., Woon, W.L.: Automatic rooftop detection using a two-stage classification. In: UKSim-AMSS 16th International Conference on Modelling and Simulation - UKSIM 2014, 26–28 March 2014. IEEE, Cambridge (2014)
9. Baluyan, H., Joshi, B., Al Hinai, A., Woon, W.L.: Novel approach for rooftop detection using support vector machine. ISRN Mach. Vis. **2013**(Article ID 819768), 1–11 (2013)
10. Secord, J., Zakhor, A.: Tree detection in urban regions using aerial liDAR and image data. IEEE Geosci. Remote Sens. Lett. **4**(2), 196–200 (2007)
11. Li, P., Song, B., Xu, H.: Urban building damage detection from very high resolution imagery by one-class SVM and shadow information. In: 2011 IEEE International Geoscience and Remote Sensing Symposium (IGARSS), pp. 1409–1412 (2011)
12. Paidipati, J., Frantzis, L., Sawyer, H., Kurrasch, A.: Rooftop Photovoltaics Market Penetration Scenarios. National Renewable Energy Laboratory, Golden (2008)
13. Al-Badi, A.H., Albadi, M.H.: Domestic solar water heating system in Oman: current status and future prospects. Renew. Sustain. Energy Rev. **16**(8), 5727–5731 (2012)
14. Noman, M.: Lecture module of managerial economics. School of Business Studies, Southeast University (2006)
15. Chen, C.J.: Physics of Solar Energy. Wiley, Hoboken (2011)
16. Komoto, K.: Energy from the desert: very large scale photovoltaic systems: socioeconomic, financial, technical and environmental aspects. Earthscan (2009)
17. Rehman, S., Bader, M.A., Al-Moallem, S.A.: Cost of solar energy generated using PV panels. Renew. Sustain. Energy Rev. **11**(8), 1843–1857 (2007)

Author Index

Printed in the United States
By Bookmasters